CAMBRIDGE LIBRARY COLLECTION

Books of enduring scholarly value

Life Sciences

Until the nineteenth century, the various subjects now known as the life sciences were regarded either as arcane studies which had little impact on ordinary daily life, or as a genteel hobby for the leisured classes. The increasing academic rigour and systematisation brought to the study of botany, zoology and other disciplines, and their adoption in university curricula, are reflected in the books reissued in this series.

The Natural History of Birds

Georges-Louis Leclerc, Comte de Buffon (1707–88), was a French mathematician who was considered one of the leading naturalists of the Enlightenment. An acquaintance of Voltaire and other intellectuals, he work as Keeper at the Jardin du Roi from 1739, and this inspired him to research and publish a vast encyclopaedia and survey of natural history, the ground-breaking *Histoire Naturelle*, which he published in forty-four volumes between 1749 and 1804. These volumes, first published between 1770 and 1783 and translated into English in 1793, contain Buffon's survey and descriptions of birds from the *Histoire Naturelle*. Based on recorded observations of birds both in France and in other countries, these volumes provide detailed descriptions of various bird species, their habitats and behaviours and were the first publications to present a comprehensive account of eighteenth-century ornithology. Volume 2 covers wild and domestic fowl and pigeons.

Cambridge University Press has long been a pioneer in the reissuing of out-of-print titles from its own backlist, producing digital reprints of books that are still sought after by scholars and students but could not be reprinted economically using traditional technology. The Cambridge Library Collection extends this activity to a wider range of books which are still of importance to researchers and professionals, either for the source material they contain, or as landmarks in the history of their academic discipline.

Drawing from the world-renowned collections in the Cambridge University Library, and guided by the advice of experts in each subject area, Cambridge University Press is using state-of-the-art scanning machines in its own Printing House to capture the content of each book selected for inclusion. The files are processed to give a consistently clear, crisp image, and the books finished to the high quality standard for which the Press is recognised around the world. The latest print-on-demand technology ensures that the books will remain available indefinitely, and that orders for single or multiple copies can quickly be supplied.

The Cambridge Library Collection will bring back to life books of enduring scholarly value (including out-of-copyright works originally issued by other publishers) across a wide range of disciplines in the humanities and social sciences and in science and technology.

The Natural History of Birds

From the French of the Count de Buffon

VOLUME 2

COMTE DE BUFFON
WILLIAM SMELLIE

CAMBRIDGE
UNIVERSITY PRESS

CAMBRIDGE UNIVERSITY PRESS

Cambridge, New York, Melbourne, Madrid, Cape Town, Singapore,
São Paolo, Delhi, Dubai, Tokyo, Mexico City

Published in the United States of America by Cambridge University Press, New York

www.cambridge.org
Information on this title: www.cambridge.org/9781108022996

This edition first published 1793
This digitally printed version 2010

ISBN 978-1-108-02299-6 Paperback

THE

NATURAL HISTORY

OF

B I R D S.

FROM THE FRENCH OF THE

COUNT DE BUFFON.

ILLUSTRATED WITH ENGRAVINGS;

AND A

PREFACE, NOTES, AND ADDITIONS,

BY THE TRANSLATOR.

IN NINE VOLUMES.

VOL. II.

LONDON:

PRINTED FOR A. STRAHAN, AND T. CADELL IN THE STRAND;
AND J. MURRAY, Nº 32, FLEET-STREET.

MDCCXCIII.

CONTENTS

OF THE

SECOND VOLUME.

A 2

10. The

CONTENTS.

3

The

CONTENTS.

4. The

CONTENTS.

C O N T E N T S.

T H E

THE BUSTARD.

THE

NATURAL HISTORY

OF

BIRDS.

GREAT BUSTARD.

L'Outarde, Buff.
*Otis * tarda*, All the Naturalifts.

WHEN we undertake to clear up the hiftory of an animal, our firft bufinefs is
to examine, with a critical eye, the various
names which it has received in different languages, and at different times; and to endeavour,
as much as poffible, to diftinguifh the feveral fpecies to which thefe have been applied. This is
the only way of reaping benefit from the knowledge acquired by the ancients, and of con-
necting it ufefully with the difcoveries of the
moderns; and confequently, the only way of

---

* In Greek, Ω]ις: In Latin, *Avis tarda*; or Slow bird: and
from this the Italian name *ftarda* is evidently formed. And may
not the old French term *biftarde*, and the Englifh *buftard*, be only
a corruption of *avis tarda*? The German appellation *trappe*, is of
the fame origin with the Englifh verb *to trape*, and alludes to its
heavy fluggifh pace.

moderns;

making real progreſs in Natural Hiſtory. For
how could, I ſhall not ſay one man, but a whole
generation, or even a ſucceſſion of generations,
complete the hiſtory of a ſingle animal? Almoſt all animals fear man and fly from him.
The character of ſupremacy, which the Moſt
High has ſtamped on his brow, inſpires them
with terror rather than reſpect. They ſhrink
from his eye; they ſuſpect his ſnares, and
they dread his arms. Even thoſe that are
able to defend themſelves by their ſtrength,
or reſiſt an attack by their bulk, retire into deſerts for which we diſdain to contend,
or entrench in the faſtneſſes of impenetrable
foreſts. The ſmall animals, ſecure in eſcaping
our vigilance by their diminutive ſize, and em-
boldened by their weakneſs itſelf, live in the
midſt of us, in ſpite of our endeavours to extirpate them, feed at our expence, and ſometimes
even prey on our own ſubſtance, though not on
that account better known. Among the great
number of intermediate claſſes included between
theſe two extremes, ſome dig for themſelves
ſubterraneous retreats, ſome plunge into the
depths of the ocean, others diſappear in the
aërial expanſe, but all of them fly from the
tyrant of Nature. How then is it poſſible, in a
ſhort ſpace of time to view all the animals in
all the ſituations neceſſary for diſcovering completely their inſtincts, their diſpoſitions, their
habits, and in a word, the principal facts of
their

their hiftory. It is well to collect at great expence numerous feries of thefe animals, to preferve carefully their external coat, to add their
fkeletons artfully combined, to give each individual its proper attitude and native air, but all
this only reprefents the furface of nature dead
and inanimate. If fome monarch would adopt
the truly grand idea, of contributing to the ad-
vancement of this beautiful part of fcience, by
forming vaft collections, and affembling, under
the eyes of obfervers, a great number of living
fpecies, we fhould ftill acquire but imperfect
ideas. Moft animals, intimidated by the prefence of man, teafed with his obfervations, and
further tormented by the uneafinefs infeparable
from captivity, would exhibit manners that are
altered, conftrained, and hardly worthy the attention of a philofopher, who admires Nature
only when free, independent, or even wild.

To ftudy animals with accuracy then, we
ought to obferve them in the favage ftate, to accompany them into the retreats which they have
chofen for themfelves, to follow them into the deep
caverns, to attend them on the frightful precipices,
where they enjoy unbounded liberty. Nor fhould
we be perceived by them while we contemplate
their habits; for the eye of an obferver, if not
concealed from their view, would, in fome meafure, difconcert their motions. But there are
few animals, efpecially of the winged tribe, that

can

can be thus furveyed : it requires a fucceffion of
ages, and innumerable fortunate occurrences, to
afcertain all the neceffary facts ; and it needs
the clofeft attention to refer each obfervation to
its proper fubject, and confequently to avoid the
confufion of names. Without thefe precautions
the moft profound ignorance fhould be preferred
to a pretended fcience, which at bottom is but a
web of uncertainty and error.   The Great Buftard is a ftriking inftance.   The Greeks named
it *Otis ;* and Ariftotle mentions it by this name
in three places * ; and his defcription perfectly
agrees with our Great Buftard.   But the Latins,
deceived probably by the refemblance of the
words, confounded it with *otus*, which is a nocturnal bird.   Pliny, after properly faying that
the bird named *otis* by the Greeks, is called *avis
tarda* in Spain, which character applies to the
Great Buftard, fubjoins, that its flefh has a
rank tafte †, which agrees with the *otus*, according to Ariftotle and to fact, but has no
reference to the Great Buftard ; and this miftake can be the more eafily fuppofed, fince

* Hift. Anim. lib. ii. 17.; lib. v. 6. ; lib. ix. 33.

† Pliny's words are : " Proximæ eis funt, quas Hifpania aves
" tardas appellat, Græcia otidas, damnatas cibis. Emiffa enim
" offibus medulla, odoris tædium fequitur." Next to thefe (he
was fpeaking of the black grous) we may rank what are termed
in Spain the *flow birds*, and in Greece, the *otides*, which are rejected as food ; for as foon as the marrow is detached from the
bones, a loathfome fmell is exhaled. Lib. x. 22.

Pliny,

Pliny, in the following chapter, evidently confounds the *otis* with the *otus*; that is, the Great Bustard with the Eared Owl.

Alexander the Myndian, as quoted by Athenæus *, falls into the same error, ascribing to the *otus* or *otis*, which he takes for the same individual bird, the circumstance of having hairy feet; which is true of the *otus*, or eared owl; in which, as in most of the nocturnal birds, the legs and feet are covered with hair, or rather clothed to the nails with feathers, that are parted into threads; and not to the *otis*, which is our Great Bustard; and in which, not only the foot, but the lower part of the leg, immediately over the tarsus, is quite bare.

Sigismundus Gelenius, having found in Hesychius the name of Ραφος, the meaning of which was not ascertained, has bestowed it, from mere fancy, on the Great Bustard †; and since his time, Mæhring and Brisson have, without assigning their reasons, applied it to the Dodo.

The modern Jews have arbitrarily taken the Hebrew word *anapha*, which denoted a kind of kite, to signify the Great Bustard ‡.

Brisson gives the word Ωτις for the Greek name of the Great Bustard, according to Belon; but afterwards adopts οτιδα from Aldrovandus. He does not advert that οτιδα is the accusative

---

* Hist. Nat. lib. ix.   † In Lexico Symphono.
‡ Paul Faugius, *apud Gesnerum.*

of

of ωτις, and confequently is the fame individual name. It is juft as if he had faid, that fome call it *tarda*, and others *tardam*.

Schwenckfeld pretends that the *tetrix*, noticed by Ariftotle *, and which was the *ourax* of the Athenians, is alfo our Great Buftard. But what little Ariftotle mentions with refpect to the *tetrix*, does not apply to the Great Buftard. The *tetrix* builds its neft among low plants, and the Great Buftard among growing corn; which Ariftotle probably did not mean to include in the general expreffion, " low plants." Secondly, This great philofopher explains himfelf in this manner: " The birds which fly little, as the " partridges and quails, do not conftruct nefts, " but lay their eggs on the ground, on fmall " heaps of leaves which they gather; the lark " and *tetrix* do the fame." The leaft attention to this paffage will convince us, that it alludes to thofe tardy birds which fly little; and that the lark and *tetrix* are mentioned, becaufe they neftle on the ground like thefe, though apparently more agile, fince the lark is of the number. If Ariftotle had meant our Great Buftard by the name *tetrix*, he would certainly have ranged it as a fluggifh bird with the partridges and quails, and not with the larks, which, from their lofty flight, have merited, according to Schwenckfeld himfelf, the epithet of *cælipetes*.

* Hift. Anim. lib. vi. 1.

Longolius

Longolius * and Gefner † are both of opinion, that the *tetrax* of the poet Nemefianus is nothing but the Great Buftard; and it muft be allowed that thefe nearly refemble each other in fize ‡ and in plumage §. But thefe analogies are not fufficient to fix the identity of the fpecies; and the lefs fo, as I find, by comparing what Nemefianus relates of his *tetrax*, with what we know of our Great Buftard, two diftinct differences: 1. The *tetrax* appears tame from ftupidity, and heedlefsly falls into the very fnare which has been laid for it ‖ ; but the Great Buftard is intimidated at the approach of man, and quickly flies out of his view ¶. 2. The *tetrax* built its neft at the foot of the Apennines; whereas Aldrovandus, who was an Italian, affures us pofitively, that the Great Buftards are never feen in Italy, except when they are driven thither by a guft of wind **. It is

---

* Dialog. de Avibus.

† De Avibus, lib. iii.

‡ " Tarpeiæ eft cuftos arcis non corpore major."—The fentinel of the Tarpeian rock (the goofe) is not larger.

§ " Perfimiles cineri dorfum maculofaque terga
  " Inficiunt pullæ cacabantis imagine notæ."—
Afh-coloured marks ftain the fhoulders (perhaps the neck) and fpeckled back, as in the partridge.

‖ " Cum pedicas necti fibi contemplaverit adftans
  " Immemor ipfe fui tamen in difpendia currit."

¶ " Neque hominem ad fe appropinquantem fuftinent, fed " cum eum longinquo cernunt ftatim fugam capeffunt."
                                        WILLOUGHBY.

** Italia noftra has aves nifi forte ventorum turbine advectas non habet. ALDROV. tom. ii.

true,

true, indeed, that Willoughby fufpeôts they are
not rare in that country; becaufe, when he
paffed through Modena, he faw one in the market. But I fhould conceive that a fingle Great
Buftard brought to market in fuch a city as Mo-
dena, agrees better with the affertion of Aldrovandus than with the conjecture of Willoughby.

Perrault imputes to Ariftotle the ftory that the
otis of Scythia does not fit on its eggs like other
birds, but covers them with a hare's or fox's fkin,
concealing them at the root of a tree, on whofe
top it is perched. Yet Ariftotle does not apply
this at all to the Great Buftard, but only to a
certain Scythian bird, probably a bird of prey;
which could tear off the fkins of hares and foxes,
and which was only of the fize of a Buftard, as
Pliny and Gaza * tranflate it; befides, however
little Ariftotle was acquainted with the Buftard,
he could not fail to know that it never perches.

The compounded name *trapp-gantz*, which the
Germans have beftowed on this bird, has given
rife to other miftakes. *Trappen* fignifies to walk;
and cuftom has connected to its derivatives the
acceffory idea of tardinefs, in the fame manner
as in the cafe of the Latin word *gradatim* and
the Italian *andante*; and hence the epithet
trapp can, with propriety, be applied to the
Buftard, which, when not purfued, walks flow-

---

* In Scythis avis magnitudine *otidis* binos parit, in leporina
pelle femper in cacuminibus ramorum fufpenfa.

Hift. Nat. lib. x. 33.

ly

ly and heavily. The application would ſtill be
juſt, though we did not affix the notion of ſlug-
giſhneſs; ſince to deſcribe a bird with the habit
of walking, contains an implication that it ſeldom
flies.

With reſpect to the word *gantz*, it may admit
of a double acceptation. Here it ought perhaps
to be written, as I have done, with a final *z* ;
and then it ſignifies *much*, and marks the ſuper-
lative : but if it be written *gans* with an *s*, it
means a *gooſe*. Some authors, taking the word
in the laſt ſenſe, have tranſlated it by the Latin
anſer trappus, and miſled by this interpretation,
have alleged that the Great Buſtard is an aquatic bird which delights in marſhes *. Aldrovandus himſelf, though informed by a Dutch phyſician of the ambiguity of the word, and though
inclined to give it the ſame meaning that I do,
has yet made Belon ſay, in his Latin tranſlation
of the paſſage, that the Great Buſtard is fond of
wet ſituations ; and yet that naturaliſt affirms directly the contrary †. This error has produced
another; and they have applied the name of *Great
Buſtard* to a bird that is really aquatic, to the
black and white gooſe which is found in Canada,

* Sylvaticus *apud Geſnerum.*

† " The nature of the Buſtard is to live in ſpacious plains, like
" the oſtrich, avoiding water above all things. It does not haunt
" wet places, ſince it remains among the ridges after rain, or it
" viſits the pools only to drink."

and

and in feveral parts of North America *. It was undoubtedly from the fame miftake, that Gefner received the figure of a palmipede bird from Scotland by the name of *Guftard*, which is in that country the real name of Great Buftard †, and which Gefner derives from *tarde*, flow, and *gufs* or *goofe*, which has the fame fignification in Dutch and Englifh. Here then is a bird which is entirely confined to the land, converted into an aquatic bird; and this ftrange metamorphofis has been occafioned by the equivocal meaning of words alone. Thofe who have ventured to juftify or palliate the name of *anfer trappus*, or *trapp gans*, have been obliged to fay, fome of them, that thefe fly in flocks like the geefe ‡; others, that they are of the fame fize ‖; as if thefe circumftances were fufficient to difcriminate a fpecies. For the fame reafon, the vultures and wood-grous might be claffed together. But I need not infift on an abfurdity; I haften to clofe this lift of errors and this criticifm, which may already be confidered as rather tedious, though I am convinced that it is neceffary.

Belon pretends that the *tetras alter* of Pliny § was the Great Buftard; but there is no foundation for this opinion, fince Pliny mentions the

---

* Charlevoix, Lade, Theodat, and the Lettres Edifiantes.
† *Guftard* in old Scotch, is the fame it would feem as *buftard*, and was probably a corruption of that word.
‡ Longolius, *apud Gefnerum*.   ‖ Frifch.
§ Hift. Nat. lib. x. 22.

avis

avis tarda in the fame place. It is true, that Belon, fupporting this error by another, afferts, that the *avis tarda* of the Spaniards and the *otis* of the Greeks mean the owl. But he ought to have proved: 1. That the Great Buftard inhabits lofty mountains, as Pliny affirms of the *tetras alter*, *(gignunt eos Alpes,)* which contradicts the affertions of all the naturalifts with refpect to this bird, except Barrere *. 2. That the owl, and not the Great Buftard, has really been known in Spain by the name of *avis tarda*, and in Greek by that of *otis;* but this is totally inadmiffible, as it is contrary to the teftimony of almoft all writers. What may have deceived Belon is, that Pliny mentions his fecond *tetras* as one of the largeft birds after the oftrich, which, according to Belon, is true only of the Great Buftard. But we fhall find in the fequel, that the wood grous fometimes exceeds in bulk the Great Buftard; and fince Pliny fubjoins, that the flefh of this *avis tarda* has a rank tafte, which correfponds much better with the *otus*, the long-eared owl, than with the *otis*, the Great Buftard, Belon fhould have fufpected that the naturalift confounds here the *otis* with the *otus*, as I have before remarked; and that he afcribes to the fame fpecies the qualities of two fpecies widely different from each

* Barrere admits two forts of Buftards in Europe; but he is the only perfon that has reprefented them as inhabitants of the Pyrennees. The author was born at Rouffillon, and referred to his native mountains all the animals bred in the adjacent provinces.

other,

other, though expreffed in his compilation by nearly fimilar names ; but he was not entitled to conclude that the *avis tarda* was really the long-eared owl.

The fame Belon would believe, that his *ædicnemus* was an *oftardeau*, or *ftone-curlew* ; and indeed this bird has only three toes, all of them anterior, like the Great Buftard ; but its bill is widely different, the tarfus thicker, the neck fhorter, and it feems to have more analogy to the plover than to the Great Buftard. But we fhall afterwards confider this fubject more fully.

Finally, We may obferve that fome authors, deceived probably by the refemblance of words, have confounded the name *ftarda*, which in Italian fignifies a buftard, with the name *ftarna*, which in the fame language fignifies a partridge.

From thefe difcuffions we may conclude, that the *otis* of the Greeks, and not *otus*, is our Great Buftard ; that the name Ραφος has been applied to it from inattention, as it has afterwards been to the dodo ; that that of *anapha*, given by the modern Jews, belonged formerly to the kite ; that the *avis tarda* of Pliny, or rather of the Spaniards in the time of Pliny, was fo called on account of its flownefs, and not as Nyphus would have it, becaufe it was late before it was known at Rome *(tardus)* ; that it is neither the *tetrix* of Ariftotle, nor the *tetrax* of the poet Nemefianus, nor the Scythian bird mentioned by

Ariftotle

Ariftotle in his Hiftory of Animals, nor the *tetras
alter* of Pliny, nor an aquatic bird ; and laftly,
that it is the *ftarda*, and not the *ftarna*, of the
Italians *.

To perceive the importance of this inveftigation, we need only figure in our imaginations
the ftrange and ridiculous idea which a beginner
would form of the Great Buftard, who had collected indifcriminately and with blind confidence
all that has been afcribed to this bird by authors,
or rather to the different names by which it is

* I fhall here collect the various names beftowed on this bird by
different authors :

Otis, Tarda, Biftarda, Gefn. and Charleton.

Otis, five Tarda, Johnfton.

Otis, feu Tarda avis, Aldrov.

Otis, Græcis ; *Tarda,* Ifiodoro ; *Biftarda,* Alberto, Rzacyfki,

Otis, Tarda, Sibbaldi Scotia Illuftrata. Will. and Ray.

Tarda Recentiorum, Schwen.

Tarda, Klein.

Tarda Pyrenaica, maculis nigricantibus, marginibus pinnarum rofeis,
Barrere.

Tetrax, feu Tarax Nemefiani, Longolius.

Tetraon, Schwenck. Charleton, and Klein.

Tetrix, Ourax, Ariftotelis, Schwenck.

Erythrontaon, Olaï Magni. Schwenck, Charleton, and Klein.

Anfer trappa, Rzacyn.

In Hebrew, *Albabari,* Gefn. and Aldrov. *Anapha,* Paulus Fa-
gius.

In Greek, Ο'τὶς, Ωτὶς, Ο'υτὶς, Gefn.

In Italian, *Starad.*

In German, *Trapp,* Gefn. Rzacyfki, & Frifch ; *Acker-trapp,*
Gefn. *Trappe,* Schwenck. and Rzacyn ; *Acker-trappe,* Schwenck.

In Flemifh, *Trap-ganfz,* Gefn. *Trapp-gans,* Shwenck.

In Swedifh, *Trapp.*

In Polifh, *Drop,* or *Trop,* Rzacynfki.

diftin-

diftinguifhed in their works : at one time a diur-
nal bird, at another a nocturnal ; fometimes an
inhabitant of the mountains, at other times an
inhabitant of the plains ; fometimes a native of
Europe, at other times a native of America ;
now a land bird, then an aquatic one ; fometimes granivorous, at other times carnivorous ;
fometimes extremely large, at other times very
fmall : in a word, a monfter and a chimæra. But,
to difcriminate the true qualities, it is neceffary,
as we have done, to draw a critical comparifon
between the defcriptions of former naturalifts.

But we have dwelt long enough on words ; it
is now time to proceed to things. Gefner exults
in being the firf who perceived that the Great
Buftard might be referred to the gallinaceous clafs.
It is true indeed, that it refembles this clafs in its
bill and its weight ; but it differs in its thicknefs ; in its legs, which have three toes; in the
fhape of its tail ; in the lower part of its legs being naked ; in the great aperture of its ears; in the
beards of feathers which hang under its chin,
in place of thofe flefhy membranes with which
the gallinaceous tribes are furnifhed ; not to mention the difference of the internal ftructure.

Aldrovandus is not more fortunate in his conjectures, when he takes the frugivorous eagle,
mentioned by Ælian *, for a Buftard, becaufe of

* Lib. x. According to Ælian, this eagle was called the *Eagle
of Jupiter.* It was ftill more a frugivorous bird than the Buftard,
which eats earth-worms ; for the eagle deftroyed no living creature.

its

its fize: as if the attribute of magnitude were alone fufficient to conftitute the idea of an eagle. It appears to me much more probable, that Ælian meant the great vulture, which is a bird of prey, as well as the eagle, and even ftronger than the common eagle, and which feeds on grain in cafes of neceffity. I opened one of thefe birds which had been wounded by a fhot, and which had paffed feveral days in fields of growing corn, and I found nothing in the ftomach but a green liquor, which was evidently half-digefted herbage.

We can more eafily trace the characters of the Great Buftard in the *tetrax* of Athenæus, which is larger than the biggeft cocks, (and we know that fome of thefe are of a prodigious fize in Afia,) has only three toes on the feet, has beards hanging on each fide of the bill, a mailed plumage, a deep cry, and whofe flefh has the tafte like that of the oftrich, which refembles the Great Buftard in many other refpects *. But this *tetrax* cannot be the Great Buftard, fince, according to Athenæus, it is a bird nowhere mentioned in the writings of Ariftotle ; whereas this philofopher fpeaks of the Great Buftard in feveral places.

We might alfo fufpect with Perrault, that thofe partridges of India mentioned by Strabo as equal to the goofe in fize, are a fpecies of Buftards.

* " Otis avis fidipes eft, tribus infiftens digitis, magnitudine
" gallinacei majoris, capite oblongo, oculis amplis, roftro acuto,
" linguâ offeâ, gracili collo " GESNER.

The

The male is diftinguifhed from the female by the
colours of its plumage, which are differently
diftributed and more vivid; by thofe beards of
feathers which hang from both fides of the neck,
which it is furprifing that Perrault has not mentioned, and with which Albin has improperly
ornamented the figure of the male; by its fize,
which is almoft double that of the female, a
greater difproportion than has been remarked in
any other fpecies.

Belon, and fome others who were not acquainted with the caffowary, the touyon, the dodo, or
perhaps the great vulture, confidered the Great
Buftard as a bird of the fecond magnitude, and
as the largeft next to the oftrich. But the peli-
lican, which was not known to them, is much
larger, according to Perrault. Perhaps, however,
Belon only faw a large Buftard and a fmall pelican;
and in that cafe, his miftake will be the fame
with that of many others, the afferting with
refpeēt to fpecies, what is true but of an individual.

Edwards accufes Willoughby of being grofsly
deceived, and of drawing Albin, who copied him,
into the fame error, in afferting that the Great
Buftard is fixty inches in length, from the point
of the bill to the end of the tail. In faēt, thofe
which I have meafured were only three feet and
a half; and fuch was that of Briffon. The one
examined by Edwards, was three feet and a half
long, or three feet nine inches from the point
of

of the bill to the extremity of the tail. In the *Britiſh Zoology*, it is ſtated at four Engliſh feet. The expanſion of the wings varies more than one half in different ſubjects. It is reckoned ſeven feet four inches by Edwards, nine feet by the authors of the *Britiſh Zoology*, and four French feet by Perrault, who declares that he never examined the males, which are always larger than the females.

The weight of this bird admits of conſiderable variations; ſome are only ten pounds, others twenty-ſeven or even thirty. But it alſo varies in its proportions; and the individuals of the ſpecies ſeem not all formed after the ſame model. Perrault obſerved ſome whoſe neck was longer, and others where it was ſhorter, than the legs; ſome whoſe bill was more pointed, others whoſe ears were ſhaded with longer feathers; and all of them had a much longer neck and legs than thoſe examined by Geſner and Aldrovandus. In the ſubjects deſcribed by Edwards, there were on each ſide of the neck two naked ſpots, of a violet colour, but which appeared covered with feathers, when the neck was much extended; a circumſtance that has been remarked by no other obſervers. Finally, Klein mentions that the Great Buſtards in Poland are not exactly like to thoſe in France and in England; and indeed we find, by comparing the deſcriptions, ſome differences in the colours of the plumage, in the bill, &c.

In general, the Great Buftard is diftinguifhed
from the oftrich, the caffowary, the touyon, and
the dodo, by the circumftance, that its wings,
though little proportioned to its mafs, are yet
able to raife it from the ground, and fupport it
for fome time in the air; whereas thefe four
birds are totally incapable of flying. It is alfo
difcriminated from all the others by its fize; its
feet, which are furnifhed with three toes, that
are parted and without membranes; its bill refembling that of the dodo; its rofe-coloured
down, and the nakednefs of the lower part of
the feet; not indeed by any one of thefe characters, but by the conjunction of them all.

The wing confifts of twenty-fix quills, according to Briffon; and of thirty-two or thirty-
three, according to Edwards, who perhaps includes thofe of the falfe wing. The only thing
I have to remark on thefe quills, and which can
hardly be perceived from the infpecting of the
figure, is that, at the third, fourth, fifth, and
fixth feathers of each wing, the exterior webs
become at once fhorter, and confequently thefe
quills are narrower, where they project from under the coverts.

The quills of the tail amount to twenty, and
the two middle ones differ from all the reft.

Perrault * imputes to Belon the afferting that
the upper part of the wings of the Great Buftard

* Memoires pour fervir a l'Hiftoire des Animaux.

is white, contrary to the obfervations of the academicians, and to what is commonly remarked
in thefe birds, in which there is more white on
the belly and the under part of the body, and
more brown and other colours on the back and
wings. But I am inclined to think, that Belon
may be eafily juftified on this head; for he fays
exactly what the academicians do, that *the Great
Buftard is white under the belly and below the
wings;* and when he defcribes the upper part of
the wings as black, he undoubtedly means thofe
quills of the wing which are next the body, and
which are really over the wing when it is clofed
and the bird in an erect pofture. But in this
fenfe the affertion is true, and conformable to
the defcription of Edwards, where the twenty-
fixth quill, and thofe that follow, inclufive to the
thirtieth, are perfectly white.

Perrault has made a more accurate obfervation.
It is, that fome feathers of the Great Buftard are
covered with down, not only at their bafe, but
even at their extremity; fo that the middle of
the feather, which confifts of clofe connected
webs, is fituated between two parts, where there
is no down. But what is very remarkable, the
down at the origin of all thefe feathers, except
the quills at the end of the wing, is of a bright red,
approaching to rofe colour, which is a character
common to the Great and the Small Buftard.
The end of the quill is alfo of the fame colour.

The

The foot, or rather the *tarfus*, and the lower part of the foot, which articulates with the tarfus, are covered with very fmall fcales, thofe of the toes being long narrow tablets; they are all of a grey colour, and fheathed with a cuticle, which it cafts like the flough of a ferpent.

The nails are fhort and convex, both above and below, like thofe of the eagle, termed *Haliætos* by Belon; fo that a fection perpendicular to their axis, would be nearly circular.

Salerne was miftaken, in afferting, that the Great Buftard, on the contrary, had nails concave below.

Under the feet, we can perceive behind a callous prominence, which ferves inftead of a heel *.

The breaft is thick and round †; the width of the aperture of the ears is probably fubject to variations; for Belon found, that it was larger in the Great Buftard than in any other land bird ‡, while the academicians could perceive nothing unufual. Thefe apertures are concealed under the feathers; and internally we difcover two ducts, one of which may be traced into the bill, and the other leads to the brain.

In the palate and the lower part of the bill, there are fituated, under the membrane that co-

* Belon and Gefner.   † Belon.
‡ " One may eafily put the tip of the finger in the paffage."
                                        BELON.

vers

vers thefe parts, feveral glands which open into the cavity of the bill by very diftinct mouths *.

The tongue is flefhy without; and within it is furnifhed with a cartilaginous nut, fixed to the *os hyoïdes*, as in moft birds; its fides are befet with points, that confift of a fubftance intermediate between membrane and cartilage. The tongue is hard, and terminates in a point; but is not forked, as alleged by Linnæus, who, with others, has undoubtedly been mifled by a wrong punctuation in Aldrovandus †.

Under the tongue, appears a kind of fac, containing about feven Englifh pints, and which Dr. Douglas, who firft difcovered it, fuppofes to be a refervoir, which the bird fills with water, to ferve as a fupply, while it wanders in the midft of thofe vaft and parched plains which it naturally prefers. But this fingular refervoir is peculiar to the male ‡, and I fufpect has given rife to a miftake of Ariftotle's. That great naturalift afferts, that the œfophagus of the Great Buftard is wide through its whole length ‖; but the moderns, and particularly the academicians, have obferved,

* Belon.

† *Lingua ferrata, utrimque acuta.* The *utrimque* ought to be feparated from *acuta*, and joined to *ferrata*. It is only a tranflation of Belon: " Sa langue eft denteleè de chaque côté, pointue " et dure par le bout."—Its tongue is indented on each fide *(utrimque)*, pointed and hard at the tip.

‡ Edwards.

‖ Hift. Anim. lib. ii. cap. ult.

that

that it enlarges only as it approaches the gizzard*. Both thefe affertions, which feem to be contradictory, may yet be reconciled, if we fuppofe that Ariftotle, or the obfervers who were employed to collect the facts for the compofition of his Hiftory of Animals, had miftaken for the œfophagus that bag, or refervoir, which is really very broad through its whole extent.

The true œfophagus, where it expands, is befet with glands regularly arranged. The gizzard, which comes next, (for there is no craw,) is about four inches long and three inches broad: it is as hard as that of ordinary hens; which is not owing, as in thefe, to the thicknefs of the flefhy part, which is here very thin, but to the internal membrane, which is extremely hard and thick, and folded and interwoven in various directions, fo as to increafe much the bulk of the gizzard.

This internal membrane appears not to be continuous, but only connected clofely to that of the œfophagus. Further, this is white, while the internal membrane of the gizzard is yellow like gold †.

The length of the inteftines is about four feet, exclufive of the *cæcum;* the internal coat of the *ileon* is ftriped with longitudinal folds, and marked at its end with fome tranfverfe wrinkles ‡.

* Gefner, Aldrovandus, and Perrault.
† Perrault, partie ii.   ‡ Ibidem.

The

The two *cæca* take their rife about feven inches from the *anus*, and ftretch forwards. According to Gefner, they are unequal in all their dimenfions; the narroweft is the longeft, and bears to the others the ratio of fix to five. Perrault fays only, that the right one, which meafures about a foot, is a little longer than the left.

Near an inch from the *anus*, the inteftine contra&s and then expands, forming a bag, which could admit an egg, and into which are inferted the ureters and the *vas deferens*. This inteftinal bag, called Fabricius's purfe *, has alfo its *cæcum*, two inches long and three inches broad; and the hole by which they communicate is covered by a fold of the internal membrane, which ferves for a valve †.

It follows from thefe obfervations, that the Great Buftard, far from having feveral ftomachs, and a great extent of inteftines, like the ruminating animals, has, on the contrary, a very fhort and narrow alimentary canal, and which is furnifhed with only a fingle ventricle. The opinion of thofe, therefore, who pretend that this bird ruminates, would be refuted by this circumftance alone ‡. Nor can we believe with Albert, that the Great Buftard is carnivorous, that it feeds on dead bodies, and even wages

---

* From the name of *Fabricius of Aquapendente*, who firft obferved it.

† Perrault.   ‡ Athenæus, Euftachius.

war againſt the feeble kinds of game; and that it never eats herbage or grain but in caſes of extreme want: far leſs ought we to conclude from theſe ſuppoſitions, that the bill and claws are hooked. Theſe errors, collected by Albert from a paſſage of Ariſtotle which is miſunderſtood, have been admitted by Geſner, with ſome modifications, but rejected by all the other naturaliſts *.

The Great Buſtard is a granivorous bird; it lives on herbs, grain, and every kind of ſeed; on the leaves of coleworts, of dandelions, of turnips, of mouſe-ear, of vetches, of ſmallage, of carrots, and even on hay, and on thoſe large worms which, during the ſummer, ſwarm before ſun-riſe on downs. In the depth of winter, and when the ground is covered with ſnow, they feed on the bark of trees; and at all times, they ſwallow ſmall ſtones, or even bits of metal, like the oſtrich. The academicians, on opening the ſtomach of one of the Great Buſtards which they obſerved, found it filled partly with ſtones, ſome of which were of the ſize of a nut, and partly with *doubloons*, to the number of ninety, all worn and poliſhed where they were expoſed to the attrition, but without the leaſt appearance of eroſion.

Willoughby found in the ſtomach of theſe birds, which were killed in the harveſt ſeaſon,

* Pennant, and others

three

three or four grains of barley, with a large quantity of hemlock feed; which indicates a decided preference, and fhews that thefe feeds would make the beft bait for enfnaring them.

The liver is very large; the gall-bladder, the pancreas, the number of pancreatic ducts, their infertion, and that of the hepatic and cyftic ducts, are liable to fome variation in different fubjects.

The tefticles are fhaped like a fmall white al-mond, and pretty firm; the *vas deferens* is in-ferted in the lower part of the fac of the *rectum*, as I have already mentioned; and, on the upper margin of the *anus*, we find a fmall appendix, which fupplies the place of a yard.

To thefe anatomical obfervations, Perrault adds this remark: That among all the fubjects diffected by the academicians, not a fingle fe-male occurred; but we have already anticipated, at the article of the oftrich, what reflections we fhould here make.

In the pairing feafon, the male ftruts round the female, and fpreads his tail into a fort of wheel *.

The eggs are not fo large as thofe of a goofe; they are of a pale olive brown, fprinkled with fmall dark fpots, in which refpect their colour bears a great refemblance to that of the plu-mage.

* Klein and Gefner.

This

This bird does not build any neft, but only fcrapes a hole in the ground *, and drops into it two eggs, which it hatches for thirty days, as ufual with large birds, according to Ariftotle †. When the anxious mother dreads the vifits of the fportfmen, fhe takes her eggs under her wings, (it is not defcribed how,) and tranfports them to a fafe place ‡. She commonly choofes fields of corn in the ear, from an inftinct which prompts all animals to bring forth their young in fituations that fupply the proper food. Klein pretends, that fhe prefers oats as having the fhorteft ftalks, and that while fhe fits on her eggs, her head is fo elevated as to glance along the plain and notice what is going forward. But this affertion agrees neither with the general opinion of naturalifts, nor with the inftinct of the Great Buftard, which, as it is wild and timid, muft feek for fafety rather by concealing itfelf in tall corn, than by over-topping it, in order to obferve the fportfmen at a diftance, and incur the danger of being itfelf difcovered.

She fometimes leaves her eggs in queft of food, and if, during her fhort abfence, one handle or even breathe on them, it is faid that fhe perceives it on her return, and abandons them.

The Great Buftard, though a very large bird, is exceffively timorous, and feems neither

---

confcious

confcious of its ftrength, nor animated by the proper fpirit of exerting it. Sometimes they affemble, to the number of fifty or fixty; but they gain as little confidence from their multitude, as from their ftrength or their fize; the flighteft appearance of danger, or rather the leaft novelty, alarms them; and they can hardly provide for their fafety, but by flight. Dogs they dread moft, efpecially as thefe are generally ufed to hunt them; but they are alfo afraid of the fox, the pole-cat, and every other animal, however fmall, which has courage to attack them. They fhrink from the fierce animals, and even the birds of prey. So daftardly they are, that, though only flightly hurt, they die through fear, rather than from the effect of their wounds *. Yet Klein afferts that they are fometimes irritated, and inflate a loofe fkin, which hangs below the neck. If we believe the ancients, the Great Buftard has no lefs affection to the horfe, than antipathy to the dog †. As foon as the timorous bird perceives that noble animal, it flies to meet him, and generally places itfelf under his feet ‡. If we admit this fympathy between fuch different animals, we might explain the fact, by faying, that the Great Buftard finds in horfe-dung fome grains that are half-digefted, and which prove a refource when preffed by hunger.

* Gefner.   † Oppian, *de Aucupio.*
‡ Plutarch, *de Soc. Animal.*

When

When it is hunted it runs exceedingly faſt, and ſometimes proceeds ſeveral miles without the leaſt interruption *. But as it with difficulty takes wing, and never unleſs aſſiſted carried by a favourable wind, and as it cannot perch on account of its weight, or by reaſon of the want of a hind toe, with which it might cling on a branch and ſupport itſelf; we may admit, on the teſtimony of both the ancients and moderns †, that it can be caught by grey-hounds. It is alſo chaſed by a bird of prey ‡ ; or nets are ſpread, into which it will be decoyed by leading out a horſe, or by merely diſguiſing one's ſelf in a horſe's ſkin §. Every kind of ſnare, how artleſs ſoever, muſt ſucceed, if it is true, as Ælian affirms, that in the kingdom of Pontus, the foxes attract them by lying on the ground, raiſing their tail, and moving it like the neck of a bird; the Buſtards, he ſays, miſtake this object for one of their own ſpecies, advance to it without heſitation, and become the prey of the inſidious animal. But this implies much ſubtlety in the fox, much ſtupidity in the Buſtard, and perhaps more credulity in the writer.

I have already mentioned, that theſe birds ſometimes flock together, to the number of fifty or ſixty: this happens in Great Britain, eſpe-

---

* Britiſh Zoology.
† Xenophon, Ælian, Albin, Friſch, &c.
‡ Aldrovandus.   § Athenæus.

cially

cially in autumn; they fpread over the turnip-
fields and commit great havock *. In France,
they are obferved to arrive and retire regularly
in the fpring and autumn, but in fmaller flocks;
and they feldom halt, except on the moft elevated fpots. They have alfo been remarked on
their paffage through Burgundy, Champagne,
and Lorraine.

The Great Buftard is found in Lybia, near
Alexandria, according to Plutarch †; in Syria,
in Greece, in Spain, in France, in the plains of
Poitou and Champagne ‡; in the open countries fituated on the eaft and fouth of Great
Britain, from Dorfetfhire to the Mers and Lo-
thians in Scotland §; in the Netherlands and
Germany ‖; in the Ukraine and Poland; where,
according to Rzacynfki, it paffes the winter in the
midft of the fnow. The authors of the Britifh
Zoology affirm, that thefe birds feldom leave
the place where they were bred, and that their
greateft excurfions never exceed twenty or thirty
miles; but Aldrovandus afferts that, towards

* Britifh Zoology. Longolius fays, that the gardeners
have great antipathy to the Buftard, on account of their deftroy-
ing the turnips. " Nec ullam peftem odere magis olitores,
" nam rapis ventrem fulcit, nec mediocri prædâ contentus effe
" folet." LONGOLIUS apud Aldrov.
† Unlefs the *otis* be confounded with *otus*, which happens fo
frequently.
‡ Salerne.
§ Britifh Zoology, *Aldrovandus*.
‖ Frifch fays, that the Buftard is the largeft of the native fowls
in Germany.

the

the end of autumn, they arrive in flocks in Holland, and limit their haunts to the fields remote from cities and inhabited places. Linnæus fays, that they travel into Holland and England. Ariftotle alfo mentions their migrations *; but this point requires to be elucidated by more accurate obfervations.

Aldrovandus accufes Gefner of a kind of contradiction on this fubject; that he affirms, that the Great Buftard migrates with the quails †, though he had mentioned before that they never leave Switzerland, and are fometimes caught in that country during winter ‡. But thefe affertions may be reconciled, if we admit, with the authors of the Britifh Zoology, that this bird only *flits*. Befides, thofe found in Switzerland are few and ftraggling, and fuch as by no means reprefent the fpecies ; and is there any proof that thofe which are fometimes caught at Zurich in the winter, are the fame individuals that lived in the country during the fummer?

What appears moft certain is, that the Great Buftard is but rarely found in mountainous or populous countries; as in Switzerland, Tyrol, Italy, many provinces of Spain, France, Eng-

* Hift. Anim. lib. viii.
† " Otidem de quâ fcribo avolare puto cum coturnicibus, fed " corporis gravitate impeditum, perfeverare non poffe, & m " locis proximis remanere."
‡ " Otis magna, fi ea eft quam vùlgo Trappum vocant, non " avolat nifi fallor ex noftris regionibus (& fi Helvetiæ rara eft,) " & hieme etiam interdum capitur apud nos." GESNER, *ibid.*

land,

land, and Germany; and that when it does occur, this happens generally in the winter *. But though it can live in cold countries, and, according to fome authors, is a bird of paffage, it would feem that it has never migrated into America by the north; for though the accounts of travellers are filled with Buftards found in the New Continent, it is eafy to perceive that thefe pretended Buftards are aquatic birds, as I have before remarked, and entirely different from that which we at prefent confider. Barrere mentions, indeed, in his Effay on Ornithology, a cinereous Buftard of America, which he fays he obferved; but in the firft place, it does not appear that he had feen it in America, fince he takes no notice of it in his account of Equinoctial France; in the fecond place, he is the only one, except Klein, who fpeaks of an American Buftard; and that of Klein, the

---

* " Memini ter quaterque apud nos captum, & in Rhætia circa " Curiam, Decembri & Januario menfibus, nec apud nos, nec " illic a quoquam agnitum." GESNER.

" The Buftard is feldom feen in Orleanois, and only in winter " during fnow." SALERNE, Ornithologie.

" A perfon of indifputed credit," fubjoins Salerne, " told " me, that one day, when the fields were covered with fnow, one " of his fervants found, in the morning, thirty buftards half- " frozen, which he brought into the houfe, and that they were " taken for turkies that had been fhut out, and were not difco- " vered till their warmth was recruited."

I recollect to have feen two myfelf at two different times in a part of Burgundy that is fertile in grain, but mountainous; but this was always in the winter feafon, and while fnow was lying on the ground.

macucagua

macucagua of Marcgrave, has not the characters
that belong to the genus, since there are four
toes on each foot, and the lower part of the leg
is feathered to its articulation with the *tarsus;*
it wants the tail, and bears scarcely any relation
to the Great Buſtard, unleſs that it is heavy,
and never flies or perches. With reſpect to
Barrere, his authority is not ſo great in natural
hiſtory, that his teſtimony can outweigh that
of all others. And, finally, his cinereous American Buſtard is probably the female of the
African Buſtard, which, according to Linnæus*,
is of an aſh-colour.

It will be perhaps aſked, how a bird, which,
though bulky, is furniſhed with wings, and
ſometimes makes uſe of them, has never migrated into America by the ſtraits on the
north, as many quadrupeds have done? I would
anſwer, that though it flies, this is only when it
is purſued; that it never makes a diſtant ex-
curſion, and, according to the remark of Belon,
has an averſion to water, and therefore could
never venture to croſs the wide expanſe of the
ocean; for, though the continents approach
each other towards the north, the interval is
ſtill prodigious, compared with the ſhort and
tardy flight of the Buſtard.

The Great Buſtard may then be conſidered
as a bird appropriated to the ancient continent,

* The *Otis Afra* of Linnæus.

but

but attached to no particular climate; it inhabits the burning fands of Lybia, and the frozen fhores of the Baltic, and occurs in all the intermediate countries.

Its flefh is excellent. That of the young ones, after being kept a fhort time, is remarkably delicate; and if fome writers have maintained the contrary, this arifes from their confounding *otis* with *otus*, as I have before obferved. I know not why Hippocrates forbids perfons fubject to the falling ficknefs to tafte it. Pliny recommends the fat of the Buftard to allay the pain in the breafts after child-birth. The quills of this bird, like thofe of the goofe and the fwan, are ufed for writing; and anglers are eager to fix them to their hooks, becaufe they believe that the little black fpots with which they are mottled, will appear to the fifh as fo many little flies, and attract them by this deception. [A]

[A] The fpecific character of the Great Buftard, *Otis-tarda* : —" The head and neck of the male is tufted on both fides." It is ranged in the order of the *Gallinæ*.

M

The LITTLE BUSTARD *.

La Petite Outarde, vulgairement *La Canepetiere,* Buff.
Otis-tetrax, Linn. Gmel. Mull. and Bor.
Otis Minor, Briff. Ray, and Will.
Tarda Nana, Klein.
Tetrax, Belon and Aldrov.
Gallina pratojuola, Cet.
The French Field Duck, Albin.

THIS bird is diſtinguiſhed from the Great
Buſtard only by ſome variations in the colours of its plumage, and in being much ſmaller.
Like the Great Buſtard, alſo, it has received the
epithet of duck *(cane),* though it has no analogy to that aquatic bird, and is never found
near ſtreams or marſhes. Belon pretends that
this name has been applied, becauſe it ſquats on
the ground as the ducks do in the water † ; and
Salerne imagines that it is on account of its re-
ſembling in ſome meaſure the wild duck, and
flying in the ſame manner. But theſe etymo-
logical conjectures are vague and uncertain ;
they reſt on a ſingle point of analogy, and are
inconſiſtent with each other ; and the name is
therefore apt to convey a falſe idea. The epi-

* The name given by Buffon, Pennant, Edward, and Latham.
† *Cane-terre,* changed into *canepetiere.*

thet

thet which we have adopted is not liable to the fame objections.

Belon fuppofes that this bird is the *tetrax* of Athenæus; refting his opinion on a paffage of the ancient, where it is compared, in point of fize, to the *fpermologus* *, which he takes for the *freux*, a kind of large crow; but Aldrovandus affirms, on the contrary, that the *fpermologus* is a fpecies of fparrow, and confequently cannot fignify the Little Buftard : and Willoughby even afferts that this bird had no name among the ancients.

Aldrovandus too informs us, that the fifhers at Rome gave the name of *ftella*, for what rea-fon he does not know, to a bird which at firft he took for the Little Buftard, but afterwards, on more minute infpection, he difcovered to be different. Yet, notwithftanding this exprefs de-claration, Ray and Salerne fay, that the Little Buftard and the *ftella avis* of Aldrovandus appear to be the fame fpecies, and Briffon places it without hefitation among the fynonyms; he feems even to allege that Charleton and Wil-loughby had the fame idea, though thefe authors have been very attentive not to confound the

---

* " The *tetrax*," fays Alexander Myndius, " is a bird of the bulk " of the *fpermologus*, of the colour of potters clay, variegated with " fome dirty fpots and great white lines : it lives on fruits, and " after it has young, it utters a cry that confifts of four parts." ATHENÆUS, lib. ix.

two

two kinds of birds, which it is moſt probable they had never ſeen *. On the other hand, Barrere, claſſing it with the rail, beſtows on it the name of *ortygometra melina*, and gives it a fourth toe to each foot; ſo true is it, that the multiplicity of ſyſtems, without increaſing our real knowledge, only ſerves to give birth to new errors.

This bird is a real Buſtard, as I have ſaid, but formed on a ſmaller ſcale; and for this reaſon Klein terms it *tarda nana, dwarf buſtard.* Its length, from the point of the bill to the end of the nails, is eighteen inches, or it is leſs than half that of the Great Buſtard. This meaſure will ſerve as a ſtandard of compariſon, from which all the other dimenſions may be deduced; but we muſt not conclude with Ray, that its bulk is to that of the Great Buſtard as one to two; it is as the cubes of theſe numbers, or as one to eight. It is nearly the ſize of a pheaſant †, and it has, like the Great Buſtard, only three toes on each foot. The lower part of its leg is naked, the bill is ſimilar to that of the gallinaceous tribe,

---

* Charleton makes two different ſpecies; the ninth, of his *phy-tivori*, which is the Little Buſtard; and the tenth, which is the *avis ſtella*. In the former he copies Belon, and in the latter he refers to Johnſton. Willoughby keeps the names of *ſtella* and *canepetiere* entirely diſtinct.

† "To have an idea of the Little Buſtard, conceive a quail much ſpotted, and as large as a middling pheaſant." BELON.

and

and there is a rofe-coloured down under all the feathers on the body ; but it has two *pennæ* fewer in the tail, one more in each of the wings; and when thefe are clofed, the laft ones ftretch almoft as far as the firft, or thofe moft remote from the body. Further, the male has not thofe beards of feathers as the male of the great fpecies ; and Klein adds, that its plumage is not fo beautiful as that of the female, contrary to what is moft ufually remarked in other birds. Excepting thefe flight differences, the two fpecies are per-fectly analogous ; they have the fame fhape, the fame internal difpofition of parts, the fame in-ftincts, the fame habits ; and it would feem that the fmall one was produced from the egg of the large, when it had not force fufficient to effect a complete developement.

The male is diftinguifhed from the female by a double white collar, and by fome other va-rieties in point of colour ; but the plumage on the upper part of the body is almoft the fame in both fexes, and, as Belon has remarked, is much lefs liable to vary in different individuals.

According to Salerne, they have a particular call in the love feafon, which begins in May. It is the found, *broo* or *proo*, which they repeat the whole night, and are heard at a great diftance. The males fight obftinately, and contend for the dominion of a certain tract ; one male takes a number of females under his protection, and the

place

place of their amours is trodden like a barn
floor.

The female lays, in the month of June, three,
four, or even five eggs, which are extremely
beautiful, and of a fhining green. When the
young are hatched, fhe leads them as a hen does
her chickens. They begin to fly about the middle
of Auguft; and when they hear a noife, they
lie flat on the ground, and fuffer themfelves to
be crufhed, rather than ftir from the fpot *.

The males are caught in fnares, into which
they are decoyed by a ftuffed female, whofe cry
is imitated. They are often hunted by means of
the falcon; but in general it is difficult to get
near them, for they are always on the watch on
fome rifing fpot in fields of oats; though never,
it is faid, among thofe of rye or wheat. Towards the clofe of the fummer feafon they pre-
pare to quit the country, and are then obferved
to affemble in flocks, and the young ones are no
longer diftinguifhable now from the old †.

According to Belon, they feed like thofe of the
great fpecies on herbs and grain, and alfo on
ants, beetles, and fmall flies; but Salerne maintains that they live chiefly on infects, and only

* Salerne. That writer does not quote his authorities. There
is fome reafon to fufpect that he confounds the *tetrix*, or wood-
cock, with the *tetrax*, or Little Buftard; efpecially as he is the only
naturalift who defcribes minutely the amours of the Little Buftard.
† Salerne.

eat

eat fometimes in the fpring the moft tender leaves of the fow-thiftle.

The Little Buftard is not difperfed through fo wide a range as the large fpecies. Linnæus fays, that it is found in Europe, and particularly in France. This affertion is rather vague; fince there are fome extenfive countries in Europe, and even large provinces in France, where it is un-known. We may refer the climates of Sweden and Poland to the number of fuch as are unfa-vourable to its nature; for Linnæus takes no notice of it in his *Fauna Succica*, nor Rzaczynzki in his Natural Hiftory of Poland; and Klein ne-ver faw more than one at Dantzic, and it came from the *menagerie* of the Margrave of Ba-reith.

Nor can it be more common in Germany; fince Frifch, who undertakes to defcribe and figure the birds in that country, and who is minute on the fubject of the Great Buftard, never mentions a word of this fpecies; and Scwenckfield never names it.

Gefner only inferts its name in the lift of thofe birds which he had never feen; and what in-deed fhews this is, that he fuppofes its feet are hairy as thofe of Attagas, which affords a fufpicion that it is at leaft very rare in Switzerland.

The authors of the Britifh Zoology, whofe view it was to take notice of no animal but what was Britifh, or at leaft of Britifh origin, con-ceive, that they would not have conformed to

their

their plan, if they had defcribed a Little Buftard that was killed in Cornwall ; but which they con-fider as a ftray bird, and by no means a native of Great Britain. So totally unknown is it in that country, that a fpecimen being prefented to the Royal Society, none of the members then prefent could recognife it, and they were obliged to apply to Edwards to difcover its fpecies *.

On the other hand, Belon informs us, that, in his time, neither the ambaffadors from Venice, Ferrara, and the Pope's dominions, to whom he fhewed one, nor any in their train, could decide what it was, and that fome of them even took it for a pheafant. From this circumftance he pro-perly infers that it muft be at leaft very uncom-mon in Italy ; and the conclufion is ftill very probable, though Ray, in paffing through Mo-dena, faw one in the market. We may there-fore reckon Poland, Sweden, Great Britain, Ger-many, Switzerland, and Italy, as countries where the Little Buftard is not found. It is even likely that the range is confined within narrower limits, and that France is the region peculiar to this bird, and the only climate fuited to its nature ; for the French naturalifts defcribe it the beft, and all the others, except Klein, who faw one, mere-ly copy Belon. Nor muft we conclude that the Little Buftard is equally common in every part of France ; there are large provinces in the king-dom where it is never feen. Salerne informs us,

* Edwards' Gleanings.

that it is pretty common in Beauce (where it is only a bird of paſſage); that it arrives about the middle of April, and retires on the approach of winter: he ſubjoins, that it delights in poor ſtoney lands, and from this circumſtance it derives the epithet of *canepetrace*, or *rock duck*. It alſo occurs in Berri, where it receives a ſimilar name*. It muſt be common in Maine and Normandy; ſince Belon, judging of the other provinces from theſe with which he was beſt acquainted, aſſerts, that *there is not a peaſant in the country who does not know its name.*

The Little Buſtard is naturally cunning and ſuſpicious; inſomuch that it has given riſe to a proverb. When it is apprehenſive of danger, it immediately quits the ſpot, and, keeping cloſe to the ground, flies ſwiftly 200 or 300 paces forward, and then runs ſo faſt that a man can hardly overtake it †.

The fleſh of the Little Buſtard is black, and is excellent food. Klein aſſures us, that the eggs of the female in his poſſeſſion were very palatable, and that the fleſh was better than that of the female of the black grous.

Its internal ſtructure is nearly the ſame, according to Belon, as that of the common granivorous birds. [A]

* *Canepetrotte.*   † Belon.

[A] Specific character of the Little Buſtard, *otis-tetrax*:—" Its " head and throat ſmooth." Latham adds, that " it is variegated " with black rufous and white, and the under ſurface white."

It is frequent in the ſouthern plains of Ruſſia, and even penetrates into Great Tartary; but it is never found in Siberia.

M

FOREIGN BIRDS

THAT ARE

ANALOGOUS TO THE BUSTARDS.

I.

The LOHONG, or CRESTED ARABIAN BUSTARD.—*Buff.*

Otis Arabs, Linn. Gmel. Briff. and Klein.
The Arabian Buftard, Lath. and Edw.

THE bird which the Arabians call *Lohong*, and which Edwards firft figured and defcribed, is nearly the fize of our Great Buftard, and, like it, has three toes on each foot, turned the fame way, only rather fhorter; the feet, the bill, and the neck are longer; and, on the whole, it is rather more taper fhaped.

The plumage on the upper part of the body is browner, and fimilar to that of the woodcock; or it is tawny and radiated with deep brown with white fpots, in the form of a crefcent, on its wings. The lower part of the body is white, as alfo the margin of the upper part of the wing. The crown of the head, the throat,

and

and the fore fide of the neck, are marked with tranfverfe bars of a dull brown on a cinereous ground. The lower part of the leg, the bill, and the feet, are of a bright brown, and yellowifh; the tail droops like that of the partridge, and is ftained with a crofs black bar; the great quills of the wing and the creft are alfo of the fame colour.

This creft forms a remarkable character in the Arabian Buftard; it is pointed, directed backwards, and much inclined to the horizon: from its bafe it fends off two black lines, of which the longer one paffes over the eye, and makes a kind of eye-lid; the other, which is much fhorter, ftretches under the eye, but does not reach it; the eye is black, and placed in a white fpace.

When we take a profile view of this creft at a little diftance, we might fancy that we fee ears pretty clofe to the head, and leaning backwards; and as the Arabian Buftard was undoubtedly better known to the Greeks than ours, it is probable that they named it *otis*, on account of thefe kind of ears, in the fame way that they have called the long-eared owl *otus* or *otos*, by reafon of two fimilar tufts which diftinguifh that fpecies of nocturnal birds.

An individual of this kind, which was brought from Moka, lived feveral years at London, in the poffeffion of Sir Hans Sloane; but Edwards, who has given us a coloured figure of it, has preferved no account of its difpofitions, its habits,

habits, or even of its manner of feeding; he ought at leaft to have not confounded it with the gallinaceous tribes, from which it differs fo widely, as I have fhewn in the article of the Great Buftard. [A]

[A] Linnæus charaƈterifes the Arabian Buftard *Otis Arabs*, by its " ereƈt tufted ears." It inhabits Arabia Felix, and penetrates in Afia as far as the Cafpian Sea.

II.

The AFRICAN BUSTARD.—*Buff.*

Otis Afra, Gmel.
Otis Atra, Linn.
The White-eared Buftard, Lath.

THIS is what Linnæus makes his fourth fpecies; it differs from the Arabian Buftard by the colours of its plumage, the black predominating; but the back is cinereous, and the ears white.

In the male the bill and feet are yellow, the crown of the head afh-coloured, and the exterior margin of the wings white; but the female is entirely cinereous, except the belly and thighs, which are black, as in the Indian Buftard.

This bird is found in Ethiopia, according to Linnæus; and it is extremely probable that the

one

one mentioned by the navigator Le Maire, by the name of *flying oftrich* of Senegal, is the fame ; for though the account given by him be fhort, it partly coincides and is entirely confiftent with the defcription of the naturalift. Its plumage is grey and black, its flefh delicious, and its fize is nearly the fame with that of the fwan. Our conjecture receives additional force from the tefti-mony of Adanfon; that intelligent naturalift hav-ing killed one of thefe flying oftriches at Senegal, and examined it narrowly, affures us, that, in many refpects, it is analogous to the European Buftard, but differs in the colour of its plumage, which is generally of a grey-afh in the greater length of its neck, and alfo by a kind of creft on the back of the head.

This creft is evidently what Linnæus calls the *ears*, and the grey-afh colour is exactly that of the female; and as thefe are the principal cha-racters by which the African Buftard of Linnæus and the flying oftrich of Senegal are diftinguifh-ed from the European Buftard, it would feem that we may conclude that they have a great analogy: and for the fame reafon we may apply to both what is obferved with refpect to each individual; for example, that they are nearly as large as our buftard, and have a longer neck. The laft men-tioned circumftance, noticed by Adanfon, is a point of refemblance to the Arabian Buftard, which inhabits almoft the fame climate ; and no-thing to the contrary can be inferred from the

filence

filence of Linnæus, fince he gives no meafure-
ment at all of the African Buftard. With regard
to bulk, Le Maire makes that of the flying
oftrich equal to that of the fwan ; and Adanfon
reprefents it as the fame with that of the European Buftard: fince, while he mentions that the
refemblance is complete in many refpects, and
ftates the principal differences, he omits that of
the fize ; and alfo as Ethiopia or Abyffinia,
which is the native region of the African Buftard,
and Senegal, which is that of the flying oftrich,
though widely differing in longitude, are of the
fame climate ; I conceive that there is great probability that thefe two birds belong to the fame
identical fpecies. [A]

[A] The fpecific character of the African Buftard, *Otis Afra*, is :
—" That it is black, its back cinereous, its ears white." It is called *korr-haen* by the Dutch at the Cape of Good Hope. Sparrman
fays, that it artfully conceals itfelf till one comes pretty near it,
when it fuddenly foars almoft perpendicularly aloft, with a fharp
quavering fcream, *korrh, korrh*, which gives the alarm to the animals in its neighbourhood.

III.

The CHURGE or MIDDLE INDIAN BUSTARD.—*Buff.*

Otis Bengalenfis, Gmel.
Pluvialis Bengalenfis major, Briff.
Indian Buftard, Edw. and Lath.

THIS Buftard is not only fmaller than the European, the African, or the Arabian fpecies, but it is taller and more flender. It is twenty inches high, from the crown of the head to the plane on which it ftands; its neck feems to be fhorter in proportion to its feet; but in other refpects it is entirely analogous to the Common Buftard. It has three feparate toes on each foot; the lower part of the leg is not feathered; the bill is fomewhat hooked, though more elongated. I am at a lofs to conceive why Briffon referred it to the genus of plovers.

The diftinguifhing character between the plovers and the buftards confifts, according to that naturalift, in the form of the bill; which, in the latter, is an arched cone, and in the former it is ftraight, and enlarged near the extremity. But in the Indian plover the bill is curved rather than ftraight, and not at all fwelling near the point as in the plovers; at leaft fo

it

it is reprefented in a figure of Edwards, which
Briffon allows to be exact. I may add, that
this property is more remarkable than in the
Arabian Buftard of Edwards, the accuracy of
which figure is alfo admitted by Briffon; and
yet he has not hefitated to clafs it with the
buftards.

We need only caft a glance on the figure of
the Indian Buftard, and compare it with thofe
of the plovers, to be convinced that it differs
totally in its appearance and proportions: its
neck is longer, its wings fhorter, and its fhape
more expanded; and befides, it is four times the
bulk of the largeft plover, whofe extreme length
is only fixteen inches, while that of the Indian
Buftard is twenty-fix *.

Black, fulvous, white, and grey, are the predominant colours of its plumage, as in the European Buftard; but they are differently diftributed. The black is fpread on the crown of the
head, on the neck, the thighs, and the lower
part of the body; a bright yellow occupies the
fides of the head and the circuit of the eyes; a
browner yellow, and one more fhaded with
black, ftains the back, the tail, that part of
the wings next the back, and the top of the
breaft, where it forms a broad belt on a dark

---

* This is confiftent with the meafure I have ftated above, that
it is twenty inches from the crown of the head to the plane on
which it ftands; for the bill and toes are not then taken into account.

ground;

ground; the white appears on the coverts of the wings fartheft from the back, and white mixed with black on the intermediate fpace; the deepeft gray is laid on the eye-lids, the extremity of the longeft quills of the wing*, of fome of the middle and fhorteft ones, and on fome of their coverts; laftly, the brighteft gray, which verges on white, is fpread on the bill and the feet.

This bird is a native of Bengal, where it is called *Churge*. We may remark, that the climate of Bengal is nearly the fame with that of Arabia, Abyffinia, and Senegal, where the two preceding Buftards are found; and we may term it the *Middle Buftard*, becaufe it holds the intermediate rank between the large and the fmall fpecies. [A]

---

* As in fome of the European Buftards.   PERRAULT.

[A] Specific charaĉter of the Indian Buftard, *Otis Bengalenfis*: —" It is black; the fpace about the eyes dufky; the back, the " rump, and the tail, dufky, but gloffed."

IV.

The HOUBARA, or LITTLE-CRESTED AFRICAN BUSTARD.—*Buff*.

Outis-Houbara, Gmel.
The Ruffed Buftard, Lath.

WE have found, among the Great Buftards, that fome are crefted and others not; and we fhall difcover that the fame diftinction prevails in the Little Buftards. That which the people of Barbary call *Houbara,* is actually decorated with a creft or ruff. Dr. Shaw, who gives us a figure of it, afferts pofitively, that it has the fhape and plumage of the Great Buftard, but is much fmaller, not exceeding the fize of a capon; for this fingle reafon, that intelligent traveller, who was certainly not acquainted with the little fpecies which inhabits France, finds fault with Golius for tranflating the word *Houbaary* by Buftard.

It lives like ours on vegetable fubftances and infects, and generally inhabits the borders of the defert.

Though Dr. Shaw takes no notice of the ruff in his defcription, there is one in the figure to which he refers; and it appears bending backwards and pendant. It is formed by long feathers

thers which rife from the neck; and which, as in the domeftic cock, briftle when the bird is irritated.

" It is curious," fays Dr. Shaw, " to obferve, " when it apprehends the attack of a rapacious " bird, the turnings and windings, the marches " and countermarches which it performs; in a " word, the evafions and ftratagems which it " makes to elude its enemy."

This learned traveller fubjoins, that it fur-niſhes an excellent medicine for fore-eyes; and that, for this reafon, its gall, and a certain fub-ftance found in its ftomach, are fometines fold at a very high price.   [A]

[A] Specific character of the Ruffed Buftard, *Otis-Houbara :* —" Yellowiſh, the feathers of the neck very long, whitiſh, and " ftriated with black ; the quills of the wings large and black, " and marked near the middle with a black fpot."

V. The

V.

The RHAAD, another SMALL-CRESTED AFRICAN BUSTARD.—*Buff.*

Otis-Rhaad, Gmel.
The Rhaad Buſtard, Lath.

T H E Rhaad is diſtinguiſhed from the Little
Buſtard of France by its creſt, and from the
Houbaara of Africa by the defect of the ruff.
It is however of the ſame ſize with the latter;
its head is black; its creſt deep blue; the upper
part of the body and the wings yellow, ſpotted
with brown; the tail of a brighter brown, radiated tranſverſely with black; the belly white,
and the bill ſtrong, as well as the legs.

The Little Rhaad differs from the Great one
by its ſize, (being no larger than a common hen,)
by ſome varieties in the plumage, and by the
want of a creſt. But it may ſtill poſſibly be of
the ſame ſpecies with the other, and differ only
by its ſex. My reaſons for this conjecture are
theſe: 1. It inhabits the ſame climate, and is
called by the ſame name. 2. In almoſt all birds,
except the carnivorous kinds, the male ſeems to
have more power of developement, which appears in their greater height, the ſtrength of their
muſcles, and in certain excreſcences, as fleſhy
membranes,

membranes, fpurs, &c. or by tufts, crefts, and ruffs, which proceed, as it were, from the luxuriancy of organization, and even by the brightnefs of the colours of their plumage.

At any rate, both the Great and the Little Rhaad are termed *Saf-faf*. *Rhaad* fignifies thunder in the African language, and is expref-five of the noife that thefe birds make in fpring-ing from the ground. *Saf-faf* denotes the ruftling of their wings when flying *. [A]

* Shaw's Travels.

[A] Specific charaƈter of the Rhaad :—" There is a creft on " the back of the head in the male, of fky-blue; the head " black; the upper-fide of the body and the wings yellow, fpotted " with dufky colour; the abdomen white, the tail dufkifh, with " black tranfverfe ftreaks." It is gregarious and granivorous in Arabia.

The C O C K.

Le Coq, Buff.
Phafianus **Gallus** *, Gmel.

THIS bird, though a domeftic, and the moft common of all, is ftill, perhaps, not fufficiently known. Moft perfons, if we except the few who beftow particular attention on the productions of Nature, need fome information with refpect to the peculiarities of its external form, and of its internal ftructure; its habits, original and acquired; the differences occafioned by fex, climate, or food; and concerning the various races which fooner or later have branched from the primitive ftock.

But if the Cock be too little known by the bulk of men, what embarraffment muft it give to the methodical naturalift, who is never fatisfied till he refer every object to his claffes and genera? If he adopts the number of toes as the foundation of his fyftem, he will range it with the birds that have four. But what place will

---

* In Greek, it was called Αλεκλωρ, from *α, priv.* & Λεκλρον, a couch, on account of its early crowing.

In Latin, *Gallus :* in Spanifh and Italian, *Gallo :* in German, *Han :* in Polifh, *Kur,* or *Kogut :* in Swedifh, *Hoens,* or *Tupt.*

he

THE DOMESTIC COCK.

he affign to the hen with five toes, which is
undoubtedly of the gallinaceous tribe, and of an
ancient family; fince it can be traced to the
time of Columella, who mentions it as a generous breed * ? If he forms the Cock into a feparate genus, diftinguifhed by the fingular fhape
of its tail, where will he place the Cock that has
no rump, and confequently no tail, but which
ftill belongs to the fame family? If he admits
that the legs clothed with plumage to the heels,
is a generic character, will he not be puzzled in
claffing the rough-footed Cock, which is feathered to the origin of the toes, and the Japa-
nefe Cock, which is feathered as far as the
nails? Laftly, If he would refer the gallinaceous
birds to the granivorous tribe, and infer, from
the number and ftructure of their ftomachs and
inteftines, that they were deftined to feed on
grain and vegetable fubftances, how will he account for the fondnefs which they difcover for
earth-worms and minced-meat, whether raw or
cooked? But perhaps, while he imagines that
the long inteftines and double ftomachs in poultry prove that they are granivorous birds, he
would alfo conclude, from the hooked fhape of
their bill, they are alfo vermivorous, or even
carnivorous. What abfurdities and contradictions! Such are the feeble efforts of a little

---

* " They are reckoned the moft noble which have five toes."
COLUMELLA, lib. viii. 2.

mind,

mind, which being unable to comprehend the
extenfion and grandeur of the univerfe, endeavours to confine it within the tramels of fyftem!
And to what trifling and vague fpeculations do
thefe attempts give rife? For our parts, we fhall
not attempt to connect the birds by a fcientific chain; we fhall only join thofe together
that feem the moft analogous; but we fhall
endeavour to mark their characteriftic features,
and note particularly the leading facts in their
hiftory.

The Cock is a heavy bird, whofe gait is compofed and flow.   His wings are very fhort, and
hence he flies feldom, and fometimes his fcreams
indicate the violence of the effort.   He crows
either in the night or day, but not regularly at
certain hours; and his note differs widely from
that of the female.   Some hens make a kind of
crowing, though fainter and not fo diftinctly
articulated.   He fcrapes the ground to feek his
food, and fwallows, with the grains, little pebbles, which rather affift digeftion.   He drinks,
by taking a little water into his bill, and raifing
his head at each draught.   He fleeps ofteneft
with one foot in the air *, and his head covered
by the wing on the fame fide.   In its natural
fituation, the body is nearly parallel to the

* The thigh on which the body refts is commonly more flefhy
than the other; and our epicures know well how to diftinguifh
them.

ground,

ground, and fo is the bill; the neck rifes verti-
cally, the forehead is ornamented with a red
flefhy comb, and the under-part of the bill with
a double pendant of the fame colour and fubftance; this however is neither flefh nor membrane, but of a peculiar nature, different from
every thing elfe.

In both fexes the noftrils are fituated on
either fide of the upper mandible, and the ears
on either fide of the head, and below each ear
a white piece of fkin is fpread.   The feet have
commonly four toes, fometimes five, but always
three of them placed behind.   The feathers rife
two and two from each fhaft; a remarkable
chara&er, which has been noticed by few naturalifts.   The tail is nearly ftraight, but admits
of a fmall elevation and depreffion.   In thofe
gallinaceous tribes where it is fingle, it confifts of fourteen feathers, which are parted into
two unequal planes that join at their upper margin, making an angle more or lefs acute.   But
what diftinguifhes the male is, that the two feathers in the middle of the tail are much longer
than the reft, and are bent into an arch; that the
feathers of the tail and rump are long and narrow,
and that the feet are armed with fpurs.   It is indeed true, that fome hens alfo have fpurs, but this
rarely occurs; and in fuch hens there are many
other points of refemblance to the male; their
comb and tail are arched the fame way; they
imitate the crowing of the cock, and would even

attempt

attempt to perform his office *. But we should
be miftaken, were we to infer that they are
hermaphrodites; they are unfit for procreation,
and averfe to the male embrace; we muft regard them as imperfect degenerate individuals,
wherein the fexual character is obliterated.

A good Cock is one whofe eyes fparkle with
fire, who has boldnefs in his demeanour, and
freedom in his motions, and all whofe proportions
difplay force. Such a bird would not indeed
ftrike terror into a lion, as has often been faid
and written, but would command the love of
the females, and place himfelf at the head of a
numerous flock of hens. To fpare him, he
ought not to be allowed more than twelve or
fifteen. Columella recommends that thefe fhould
not exceed five; but, though the Cock fhould
have fifty a-day, it is faid † that he would
not neglect one. Yet no one can be certain
that all his embraces are efficacious, and fufficient to fecundate the eggs of the female. His
luft feems to be as fiery as his gratifications are
frequent. In the morning, the firft thing he
does, after he is let out from his rooft, is to
tread his hens. Food feems to him only a fecondary want; and if he is deprived for fome
time of the company of his family, he makes
his addrefles to the firft female that he meets,

* Arift. Hift. Anim. lib. ix. 49.
† Aldrovandus.

though

though of a very different fpecies *, and even
courts the firft male that occurs.   The firft fact
is mentioned by Ariftotle ; the fecond is proved
by an obfervation of Edwards † ; and by a law
mentioned by Plutarch, in which it was enacted,
that a Cock convicted of this unnatural act,
fhould be burnt alive ‡.

The hens muft be felected for the Cock, if we
would have a genuine race ; but if we want to
vary and improve the fpecies, the breed muft be
croffed.   This obfervation did not efcape the ancients : Columella exprefsly mentions, that the
beft poultry is produced by the union of a Cock
of a foreign family with the ordinary hens ; and
we find in Athenæus, that this idea was improved, a cock-pheafant being given to the common hens ‖.

In every cafe we ought to chufe thofe hens
which have a lively eye, a flowing red comb, and

---

* A crofs-breed is produced between a Cock and the hen-par-
tridge, which through time grows like the female.
                                        ARISTOTLE, lib. ix. 49.

† Having fhut up three or four Cocks in a place where they
could have no commerce with any hen, they foon laid afide their
former animofity ; and, inftead of fighting, each tried the other,
though none feemed willing to fubmit.   *Preface to the Gleanings.*

‡ In his treatife on the queftion, " Whether brutes reafon ?"

‖ *De Re Ruftica,* lib. viii. 2.   Longolius defcribes the method
of affociating the cock-pheafant with common hens.   GESNER *de
Avibus.* 1 am affured that the Guinea Cock alfo treads the hens,
if educated together, but that the breed are rather barren.

have

have no fpurs. The proportions of their body
are in general more flender than the males; yet
their feathers are broader, and their legs fhorter.
Sagacious farmers prefer black hens, becaufe they
are more prolific than the white, and more eafily
efcape the piercing fight of the birds of rapine
which hover near the farm-yard.

The Cock is extremely watchful of his females, and even filled with inquietude and anxiety; he hardly ever lofes fight of them; he leads
them, defends them, and threatens them with
his menaces; collects them together when they
ftraggle, and never eats till he has the pleafure
of feeing them feeding around him. To judge
from the different inflexions of his voice, and
the various fignificant geftures which he makes,
we cannot doubt but thefe are a fpecies of language that ferves to communicate his fentiments.
When he lofes them, he utters his griefs. Though
as jealous as he is amorous, he abufes not his
wives, but turns his rage againft his rivals. When
another Cock is prefented, he allows no time for
feduction; he inftantly rufhes forward, his eyes
flafhing fire, and his feathers briftled, and makes a
furious attack on his rival, and fights obftinately
till one or the other fall, or the interloper leave
the field. The defire of poffeffion, ever exceffive,
not only prompts him to drive away every rival,
but to remove the moft inoffenfive obftacles; he
beats off and fometimes kills the chickens, that
he may enjoy the mother more at his eafe. Is
this

this appetite the fole caufe of his furious jealoufy? In the midft of a fubmiffive feraglio, how
can he apprehend any bounds to his gratification? But how ardent foever be his paffions, he
feems to be more averfe to fhare the pleafures
than eager to tafte them; and as his powers are
greater, fo his jealoufy is more excufable and
better founded than that of other fultans. Like
them alfo, he has his favourite female, whom he
courts with greater affiduity, and on whom he
beftows his favours as often nearly as on all the
reft together.

What proves that in Cocks jealoufy is a paffion founded on reflection is, that many of them
are perpetually fighting with each other in the
court-yard, while they never attack the capons,
at leaft if thefe are not in the habit of following
the hens.

Man, who is dexterous in drawing amufement from every quarter, has learnt to fet into
action that invincible antipathy which Nature
has implanted in one Cock to another. So much
have they foftered this native hatred, that the
battles of two domeftic birds have become fpec-
tacles fit to attract the curiofity of people even in
polifhed fociety; and at the fame time, thefe
have been confidered as the means of calling
forth or maintaining that precious ferocity, which
is, they fay, the fource of heroifm. Formerly,
and even at prefent in more than one country,
men of all ranks crowd to thefe grotefque combats,

bats, divide into parties, grow heated for the fortune of their favourite Cock, heighten the interest of the exhibition by the most extravagant bets; and the fate of families is decided by the last stroke of the victorious bird. Such was anciently the madness of the Rhodians, the Tangrians, and the people of Pergamus * ; and such at present is that of the Chinese †, of the inhabitants of the Philippine islands, of Java, of the isthmus of America, and of some other nations in both continents ‡.

But Cocks are not the only birds that have been thus abused: the Athenians, who allotted one day in the year ‖ to cock-fighting, employed quails likewise for the same diversion; and even at present the Chinese breed for that purpose certain small birds resembling quails or linnets. The mode of fighting varies according to the different schools where they are formed, and the different weapons, offensive or defensive, with which they are armed; but it is curious that

* Pliny, lib. x. 21.
† Gemelli Careri, Ancient Accounts of India and China.
‡ Navarette, *Defcription de la Chine.*
‖ When Themistocles was about to give battle to the Persians, obferving his troops difpirited, he pointed to two Cocks that were fighting : " See," faid he, " the unfhaken courage of thefe ani-" mals ; yet they have no other motive than the love of victory. " But you fight for your houfehold gods, the tombs of your fa-" thers, and your liberty." Thefe few words revived the courage of the army, and Themistocles gained the victory. It was in memory of this event that the Athenians inftituted a kind of feftival, which was celebrated by cock-fighting.   ÆLIAN.

the

the Rhodian Cocks, though larger, ſtronger, and
better fighters than the others, were not ſo ardent for the females, and had only three hens,
inſtead of fifteen or twenty; whether becauſe
their fire was extinguiſhed in the conſtrained ſo-
litude in which they were accuſtomed to live,
or becauſe their rage, being too often rouſed,
had ſtifled in them the ſofter paſſions, which,
however, were at firſt the principle of their courage and the ſource of their hoſtile diſpoſition.
The males of that breed were therefore leſs vigorous, and the females leſs prolific and more in-
dolent, both in laying eggs and watching their
chickens.—So ſuccefsful has Art been in degrading
Nature! and ſo unfavourable are the talents for
war to the buſineſs of propagation!

Hens need not the embrace of the Cock to procure eggs; theſe are continually detached from the
bunch in the ovarium, which grows independent
of the union with the male. As they enlarge, they
acquire maturity, ſeparate from their calyx and
pedicle, paſs through the whole length of the
oviductus, and in their road aſſimilate, by a certain power that they poſſeſs, the lymph with
which the duct is filled, and form it into their
white, their coats, and their ſhell. There they
remain till the ſenſible and elaſtic fibres being
ſtretched and ſtimulated by theſe ſubſtances, which
have now become foreign, contract and extrude
them, the large end being foremoſt according to
Ariſtotle.

8   Theſe

Thefe eggs are all that the prolific quality of the female can produce alone and unaffifted; fhe exudes an organized body, indeed, fufceptible of a kind of life, but not a living animal fimilar to the mother, and in its turn capable of continuing the race. This requires the union of the male, and intimate mixture of the feminal liquors of both fexes; but when once this has taken place, its effects are durable. Dr. Harvey obferved, that the egg of a hen, which had been feparated twenty days from the Cock, was not lefs prolific than one laid newly after treading, and that the embryo was not on that account more advanced, and required the fame length of incubation; a certain proof that heat alone cannot produce or promote the developement of the chick, but that the egg muft be formed, or at leaft placed where it can perfpire, in order that the embryo inclofed may be fufceptible of incubation, otherwife all the eggs which remain in the oviduct twenty-one days after fecundation would hatch, fince they would have the proper time and heat; and, in this cafe, hens would be fometimes oviparous, fometimes viviparous *.

The mean weight of the egg of an ordinary hen is one ounce fix grains. If we open it care-

---

* I know of no perfon, except Dr. Michael Lyzeruts, who faw a viviparous hen. But the inftances would be frequent, if heat were only required to hatch a fecundated egg. *German Ephemerides,* Dec. ii. an. 4. Append. obf. 28.

fully, we may perceive, immediately under the
fhell, a common membrane which lines the whole
of the inner cavity ; then, the external white,
which has the form of this cavity ; next, the internal white, which is rounder than the preceding ; and laftly, in the centre of this white, we
find the yolk, which is fpherical. All thefe different parts are inclofed, each in its proper membrane ; and all the membranes are connected together at the *chalazæ* *, or cords, which form the
two poles of the yolk. The little lenticular ve-
ficle, called the *cicatricula* †, appears very near
its equator, and is firmly attached to the furface ‡.

With refpect to its external form, it is too well
known to need any defcription ; but it is often
altered by accidents, which it is eafy to account

* So called from Χαλαζα, a hail-ftone, on account of the fim -
larity of appearance.

† i. e. *a little fcar*. It is a yellowifh white round fpot, and
when examined, it appears compofed of feveral different coloured
circles.

‡ Bellini, mifled by his experiments, or rather by the confe-
quences which he drew from them, fancied, and made many believe, that if frefh eggs were hardened in boiling water, the *cicatricula* left the furface of the yolk, and retired to the centre ; but
when eggs that had been fet under the hen were hardened in
the fame way, the cicatricula remained conftantly attached to the
furface. The philofophers at Turin repeated and varied the experiments, but found, that in all eggs, whether new-laid or fuch
as had undergone a partial incubation, the cicatricula continued to
adhere to the furface of the hardened yolk ; and that the white fubftance which Bellini faw at the centre was quite different, and was
occafioned by too much or too little boiling.

for

for from the hiftory of the egg itfelf and its
formation.

It is not uncommon to find two yolks included
in the fame fhell. This happens when two eggs
alike formed are detached at the fame time
from the *ovarium* and pafs together through the
oviduct, forming their white without parting, and
become invefted with the fame external coat.

If by any accident, which may eafily be fuppofed, an egg that has been fome time difengaged
from the *ovarium*, is checked in its growth, and
when formed as much as it can be, comes within
the fphere of action of another vigorous egg, it
will coalefce with it, and form an egg within
an egg *.

In the fame manner, we may conceive how a
pin, or any other fubftance, which has penetrated
as far as the oviduct, will be found inclofed within † an egg.

Some hens lay eggs that have no fhell; whether from the defect of the proper fubftance for
forming the fhell, or becaufe they are extruded
from the oviduct before their complete maturity:
thefe never produce chickens; and this happens,
it is faid, to hens that are too fat. The oppofite
circumftances occafion the eggs to have too
thick fhells, or even double fhells. Some retain the pedicle by which they are fixed to the
ovarium ; others are bent into the form of a cref-

* Collection Academique.   † Idem.

cent;

cent; others are shaped like a pear; some have
had on their shells the impressions even of a sun,
a comet *, an eclipse, or whatever has operated
powerfully on the imagination; nay, some have
appeared luminous. What has been real in the
alterations in the shape of the egg, and the marks
on the surface, must be ascribed to the different
pressures which it receives while the shell is still
soft and pliant, and yet of such a consistence as
to retain the impressions. It will be more difficult to account for the luminous appearance † of
some eggs. A German doctor observed such under a white hen which had been fecundated, he
adds, by a very vigorous Cock. We cannot de-
cently deny the possibility of the fact; but, as it is
single, it would be prudent to repeat the observation before we venture to explain it.

With regard to the pretended Cocks eggs that
have no yolk, and include, as the vulgar imagine,
a serpent ‡, they are nothing else but the imma-
ture productions of an infant hen, or the last
effort of one exhausted by excessive fecundity;
or, finally, they are imperfect eggs that have lost
their yolks in the oviduct, either from accident
or from the wrong conformation of the parts,
but that have still retained their cords or *chalazæ*,
which the lovers of the marvellous have fancied
converted into a serpent. M. de la Peyronie has

---

* Collection Academique.
† Ephemerides de Curieux de la Nature.
‡ Collection Academique.

put

put this beyond all doubt, by the diſſection of a
hen which laid ſuch eggs ; but neither M. de la
Peyronie nor Thomas Bartholin, who diſſected
theſe pretended oviparous Cocks, could diſcover
eggs, or ovaria, or any thing analogous *.

Hens lay through the whole year, except the
time of moulting, which generally laſts ſix weeks
or two months, about the end of autumn and
the beginning of the winter. This moulting is
nothing but the ſhedding of the old feathers,
which are detached like the old leaves of trees
and the antlers of ſtags, being excluded by the
growth of the new. The Cocks alſo ſuffer this
renovation ; but it is remarkable that the new
feathers ſometimes aſſume a different colour.
One of our obſervers has noticed this fact in a
hen and a cock, and every perſon may remark
it in many other kinds of birds, particularly in
thoſe that are brought from Bengal, which change
their tints at almoſt every moulting ; and, in general, the colours of the firſt feathers, in by far
the greateſt number of birds, are different from
what they afterwards become.

The ordinary fecundity of hens is limited to
the laying an egg each day. There are ſome, it is
ſaid, in Samogitia †, Malacca ‖, and other places,
that lay twice a-day. Ariſtotle mentions certain
hens of Illyria, which laid ſo often as thrice a-

---

* Collection Academique.

† Rzacynſki, Nat. Hiſt. Poland.

‡ Bontekoe, Voyage aux Indes Orientales.

day ;

day; and it is probable that thefe were the fame with the Adrian or Adriatic hens, of which he fpeaks in another place, and which were noted for their prolific quality. Some add, that there is a peculiar mode of feeding common hens, which gives them this prodigious fecundity. Heat is very favourable; hens can be brought to lay in winter by keeping them in a ftable, where there is always warm dung on which they can fit.

As foon as an egg is laid it begins to perfpire, and lofes every day fome grains of its weight by the evaporation of the more volatile juices: in proportion, it thickens, hardens, and dries; or it contracts an offenfive fmell, and becomes totally unfit for hatching. The art of long preferving eggs confifts in checking the perfpiration *, by covering the fhell completely with a coat of any kind of greafe fhortly after it is laid. By this fingle precaution we can preferve them for feveral months, and even years, in a condition for eating, and capable of being hatched, and, in a word, retaining all the properties of frefh eggs †. The inhabitants of Tonquin keep them in a kind of

---

* The *Journal Economique* for the month of March 1755 mentions three eggs, fit for eating, found in Italy, in the heart of a wall built 300 years ago. This fact is the more incredible, as a coat of mortar would not be fufficient to preferve an egg; and as the thickeft walls dry in every part, the tranfpiration through the fhell would not be prevented.

† *Pratique de l'art de faire eclore le poulets.*

pafte

pafte made of fifted afhes and brine; other Indians in oil*; varnifh is alfo proper, when the
eggs are intended for the table; but greafe is
equally fit for this purpofe, and is much better
for preferving eggs that are to be hatched, becaufe it can be more eafily feparated than the
varnifh, and the coating muft be completely detached in order that the incubation fucceed ; for
whatever obftructs the perfpiration prevents alfo
the developement of the chick †.

I have faid, that the union of the Cock was
neceffary to the fecundation of the eggs ; and
this fact is founded on long and conftant experience.   But the details of this act, fo effential
in the hiftory of animals, have been too flightly
obferved.   It is indeed known, that the male organ is double, and is only the two paps which
terminate the fpermatic veffels, where they are
inferted in the gut ; that the female *vulva* is
placed over the *anus*, and not under as in the
quadrupeds ‡ : that he advances to his female
with an oblique quickened pace, dropping his
wings, like the turkey, and even partly fpreading
his tail, uttering a certain expreffive murmur,
with a trembling motion, and with all the figns

---

* Tavernier.

† This affertion feems to require fome modification. The late
ingenious Dr. Monro of Edinburgh rubbed the obtufe ends of eighteen eggs, and yet they all fucceeded in hatching.
                              Comparative Anatomy, p. 94.

‡ Rhedi.—Collection Academique.

of

of ardent defire; that he darts upon the hen, who receives him, bending her legs, fquatting on the ground, and laying afide the two fans of feathers which compofe her tail; that then he feizes with his bill the creft or tuft on the head of the female, either by way of careffing her, or of keeping his balance; that he bends the hind part of his body, where his double yard is lodged, and applies it brifkly where the correfponding orifice is placed; that this copulation lafts the fhorter time the oftener it is repeated, and that the Cock feems to boaft of his performance by clapping his wings, and by a kind of crowing of joy or victory; that he has tefticles, and that his feminal liquor is contained, like that of the quadrupeds, in fpermatic veffels. I have afcertained, by my own obfervations, that the femen of the hen is lodged in the *cicatricula* of each egg, as that of the female quadrupeds is included in the glandular bodies of the tefticles; but I am uncertain whether the double *penis* of the Cock, or only one of them, penetrates the female orifice, and even whether there is a real infertion or only a ftrong compreffion or mere contact. It is not yet known what muft be the precife condition of an egg in order to its fecundation, nor to what diftance the male influence can extend.—In a word, notwithftanding the infinite number of experiments and obfervations that have been made on this fubject, we ftill re-

main

main unacquainted with fome of the principal
circumftances of the impregnation.

Its firft known effect is the dilation of the
cicatricula, and the formation of the chick in its
cavity; for it is this *cicatricula* which contains
the true germ, and occurs in eggs whether fertile or not, and even in thofe pretended Cocks
eggs which I have already fpoken of*; but it
is fmaller in the eggs that are not fecundated.
Malpighi, having examined fertile eggs that
were newly laid and before they were covered,
difcovered in the centre of the *cicatricula* a fpeck
fwimming in a liquor, in the midft of which he
could trace the rudiments of a chick diftinctly
formed; but the *cicatricula* of barren eggs, produced by the hen alone without the intercourfe
of the male, fhewed merely a fmall fhapelefs
globule, furnifhed with appendices filled with a
thick juice, though furrounded with feveral con-
centric circles; and he could perceive no embryo of an animal.   The intimate and complete
organization of a fhapelefs mafs is only the
inftantaneous effect of the mixture of the two
feminal liquors; it requires but a moment for
Nature to give the firft form to this tranfparent

---

* De la Peyronie obferved in one of thefe eggs a round yellow
fpot, of a line in diameter, but without any fenfible thicknefs,
fituated on the membrane that adheres to the fhell.   It is probable
that the yellow colour was, in this cafe, occafioned by the difper-
fion of the yolk, which was found in diffection; the membrane
which contained the yolk, perhaps, ftuck to that next the fhell.

glareous

glareous fubftance, and to diffufe the principle
of· life through all its points; fhe requires time
and confpiring circumftances to finifh the rude
fketch. She has entrufted this charge chiefly to
the mother, by infpiring the inftinct of covering the eggs. In moft hens this propenfity is
felt as ftrongly, and marked by as fignificant
figns as copulation, to which it fucceeds in the
order of nature, and even though it is not ex-
cited by the prefence of an egg. A hen that
has juft laid, is kindled with tranfports that are
communicated to others which are only mere
fpectators, and they all join in the repeated
clamorous burfts of joy; either becaufe the fudden ceffation of the pains of delivery is always
accompanied with a lively pleafure, or that the
mother then anticipates all the delights of progeny. Whatever be the caufe, certain it is,
that when fhe has laid twenty-five or thirty eggs,
fhe deliberately prepares to fit on them. If
they be continually taken from her, fhe will perhaps lay twice or thrice as many, and become
exhaufted by the mere excefs of fecundity. But
the time at laft comes, when fhe is driven by the
force of inftinct to feek to hatch, and expreffes
her wants by a particular fort of clucking, and
by certain unambiguous motions and attitudes.
If fhe cannot find her own eggs, fhe will readily cover thofe of any other hen, or thofe of
the female of any other fpecies, or even balls of
ftone or chalk. She will continue ftill to fit,

after

after every thing is removed, wafte herfelf in vain plaints and idle movements *. If fhe is fuccefsful in her fearch, and finds eggs that are either real or refembling fuch in a retired and convenient fpot, fhe immediately feats herfelf on them, covers them with her wings, fofters a genial warmth, and conftantly changes them gently to heat all the parts equally. She is fo intent in her occupation, as to negleƈt food and drink. One would almoft fay, that fhe perceives the importance of her employment; fhe omits no care, overlooks no precaution, to complete the exiftence of the little incipient beings, and to guard againft the dangers that threaten †. It may perhaps be worthy of remark, that the condition of a fitting hen, however infipid it appears to us, is perhaps not a tedious fituation, but a ftate of continual joy, the more delicious as it is the more choice; fo much has Nature conneƈted raptures with whatever relates to the multiplication of her creatures!

The effeƈt of incubation is confined merely to the developement of the embryo of the

---

* A hen may be put off the brooding by often dipping her pofteriors in cold water.

† Noife is fometimes injurious to the brood. A whole hatch made in a lock-fmith's fhop was attacked by vertigos.
Colleƈtion Academique.

A fingular circumftance lately came under my obfervation:—A brooding hen having perceived a hole made in one of her eggs was filled with rage; but after her paffion was fomewhat abated, fhe deliberately clofed up the wound with mud and feathers. T.

chick,

chick, which, as we have already faid, exifts already formed in the *cicatricula* of the fecundated egg. The following is nearly the order in which this developement is made, or rather as it appears to the obferver; and as I have elfewhere given a pretty full account of the facts relating to this fubject, I fhall only repeat the more important circumftances.

At the end of five or fix hours, the head of the chick is diftinctly feen joined to the dorfal fpine, fwimming in the liquor, with which the fpeck in the centre of the *cicatricula* is filled; and towards the clofe of the firft day, the head is already bent back by its enlargement.

On the fecond day are perceived the firft traces of the *vertebræ*, which are like fmall globules difpofed on the two fides of the middle of the fpine; the wings and umbilical veffels alfo begin to appear, diftinguifhed by their dull colour; the neck and breaft are unfolded, and the head conftantly increafes; the outlines of the eyes, and the three encircling coats, as well as the fpine and membranes, are now feen. The life of the fœtus is decided; the heart beats, and the blood circulates.

On the third day, the whole is more diftinct and expanded. It is remarkable, that the heart hangs out of the breaft, and beats three times in fucceffion; once, in receiving from the auricle the blood contained in the veins; a fecond time, in difcharging it into the arteries; and a third

time,

time, in forcing it into the umbilical veffels;
and this motion continues for twenty-four hours
after the embryo has been feparated from the
white of its egg. We alfo difcover the veins
and arteries on the veficles of the brain, and the
rudiments of the fpinal marrow beginning to
extend along the *vertebræ*. Laftly, we fee the
whole fœtus enveloped in a part of the fur-
rounding liquor which has acquired a greater
confiftence than the reft.

On the fourth day the eyes are confiderably
advanced; we can diftinguifh the pupil, the
cryftalline lens, and the vitreous humour. We
alfo perceive in the head five veficles filled with
a fluid which, approaching each other, and
gradually coalefcing on the following days,
form at laft the brain invefted with its coats.
The wings grow, the thighs begin to appear,
and the body to acquire bulk. On the fifth
day the whole body is covered with an unctu-
ous flefh ; the heart is confined by a very thin
membrane, which fpreads over the cheft ; and
the umbilical veffels rife out of the abdomen *.

The fixth day the fpinal marrow, being divided into two parts, continues to ftretch along
the trunk ; the liver, which was before whitifh,

* The veffels which fpread in the yolk of the egg, and which
are confequently without the abdomen, foon retire into the cavity,
according to the remark of Steno.   *Collection Academique.*

is now become of a dufky colour; the heart beats
with its two ventricles; the body of the chick
is covered with a fkin, and even the feathers
begin to fprout.

It is eafy, on the feventh day, to diftinguifh
the bill; the brain, the wings, the thighs, and
the legs, have acquired their perfect fhape; the
two ventricles of the heart appear like two
bubbles, contiguous and joined above to the
fubftance of the auricles. Two fucceffive motions are obferved in the ventricles, as well as
in the auricles, which refemble two feparate
hearts.

About the end of the ninth day the lungs
appear, and are of a whitifh colour. On the
tenth day the mufcles of the wings are completely formed, and the feathers continue to
fhoot. It is not till the eleventh day that we perceive the arteries, which were before at a diftance
from the heart, cohere to it; and this organ is
now perfect and united into two ventricles.

The following days are fpent in the farther
expanfion of the parts, which continues till the
chick breaks its fhell; and this happens commonly the twenty-firft day, fometimes the
eighteenth, and at others, the twenty-feventh.

All this train of phænomena, which prefents fo interefting a fpectacle to the obferver,
is the effect of incubation by a hen; and human induftry has found it not beneath its notice to imitate the procefs. Formerly, the rude

peafants of Egypt, and in our own times philofophers, have fucceeded in hatching eggs, as
well as the moft careful fitter, and have given
birth to amazing numbers at once. The whole
fecret confifts in keeping the eggs at a temperature which nearly correfponds to the warmth of
the hen, and in preventing every kind of humidity and pernicious exhalations, fuch as thofe of
charcoal, burning fuel, and even that of tainted
eggs. By obferving thefe two conditions, and
being attentive in repeatedly fhifting the eggs,
and varying the place of the oven or ftove where
the bafkets are placed, fo that not only each
egg, but every part of it may enjoy alike the
requifite heat, we fhall fucceed in hatching mil-
lions of chickens.

Every kind of heat is favourable; nor is the
warmth of the hen better than that of any other
animal, not even excepting man *, nor than the
folar or terreftrial fires, or the heat of a bed of
oak-bark or dung. The effential point is to be
able to regulate the heat; to increafe or diminifh at pleafure. We can always know the
degree by means of good thermometers placed
in different parts within the oven or ftove; we

---

* When Livia was pregnant, fhe cherifhed an egg in her
bofom, with a view of foretelling the fex of her expected child
from that of the chicken which would be hatched. It was a
cock, and fhe had a boy. The augurs turned the accident to
their advantage, and endeavoured to convince the incredulous of
the reality of their art. But what was better proved is, that the
heat of the human body is fufficient for the incubation of eggs.

can

can preferve the heat, by ftopping the openings
and fhutting the regifters of the lid; we can
augment it with warm afhes, if it be an oven,
or by adding wood, if a ftove; or ufing chafing-
difhes, if it be a bed; and we can diminifh it,
by opening the regifters to give accefs to the
external air, or at leaft by introducing into
the oven cold bodies, &c.

But, whatever attention be beftowed in regu-
lating the furnace, it is hardly poffible to maintain conftantly, and without interruption, the
32d * degree of heat, which is that of the hen.
Fortunately, this limit is not very determined;
and a heat varying from the 38th † to the
24th ‡ degree, is found to occafion no inconvenience. But it is to be obferved, that the
excefs is more to be dreaded than the defect,
and that a few hours at the 38th degree, or
even the 36° §, is more injurious than fome
days at 24°. And a proof that a ftill lefs heat
would occafion no inconvenience is, that a partridge's neft being difcovered in a meadow that
was mowing, they were kept in the fhade for
thirty-fix hours, (no hen being found during
that time to cover them,) and yet they all
hatched at the end of three days, except thofe
which were opened to perceive what condition
they were in. They were indeed very far ad-

* 104°, Farenheit.　　† 117°½, F.
‡ 86°, F.　　§ 113°, F.

vanced,

vanced, and it undoubtedly required a greater
degree of heat at the beginning than near the
clofe of the incubation, when the heat of the
little chick was almoſt alone ſufficient for its developement.

With regard to humidity, as it is very hurtful to the progreſs of incubation, we muſt be
furniſhed with certain means of diſcovering
whether it has penetrated into the oven, and be
able to expel it, if it has penetrated, and prevent its return.

The hygrometer, which is the ſimpleſt and
fitteſt for eſtimating the humidity of the air in
theſe ſorts of ovens, is a cold egg introduced
and kept ſome time, till it acquire a proper heat.
If, at the end of a quarter of an hour or more,
the egg is covered with a light dew, ſuch as that
formed by the breath on poliſhed glaſs, or what
falls on the outſide of a tumbler in which a freez-
ing mixture is made, this is a proof that the air
of the oven is too humid ; and the more ſo, the
longer time the moiſture takes to diſſipate. This
happens chiefly in a tan-bed, and in dung com-
poſts incloſed in a confined place. The beſt
way to remedy this inconvenience is to renew
the air, where it is cloſe, by means of currents
produced by oppoſite windows ; or, inſtead of
theſe, by fixing ventilators proportioned to the
ſpace. Sometimes the mere perſpiration of the
immenſe number of eggs occaſions an exceſs
of humidity in the oven itſelf : in this caſe
the

the baſkets with the eggs ought to be taken out, for a few minutes, every two or three days, and fanned with a hat, waving it in different directions.

But it is not enough that the moiſture which collects within the oven be expelled; we muſt prevent the entrance of humidity from without, by lining the inſide with ſheet-lead, good cement, plaſter, a proper mixture of pitch and tar, or at leaſt by ſpreading it over with ſeveral coats of oil, and allowing this to dry, and gluing on the interior ſurface ſtripes of bladder or of grey paper.

To theſe few eaſy practical operations is reduced the whole art of artificial incubation; and hence are deduced the ſtructure and dimenſions of the ovens or ſtoves, the number, ſhape, and diſtribution of the baſkets, and all the little manœuvres which the circumſtances require, or the occaſion ſuggeſts, which have been deſcribed with a profuſion of words, but which we ſhall compriſe in a few lines, without omitting any thing eſſential.

The ſimpleſt oven is a caſk lined within with glued paper, and ſtopped at the head with a cover which joins into it, and which is perforated in the middle by a large opening, that ſhuts with a grooved lid, to allow an opportunity of examining the oven. There are ſeveral other ſmall holes round this, which ſerve as regiſters to regulate the heat, and which can alſo be ſtopped with grooved covers. This caſk is buried three-fourths of its height in warm dung. In the inſide there are placed, one above another, at proper diſtances, two or three wide-ribbed baſkets, in each of which two rows of eggs are piled, taking care that the upper layer be thinner than the lower, ſo that this may be eaſily ſeen through the other. Small holes may be made, if we chuſe, in the centre of each baſket; and well-graduated thermometers ſuſpended there, and others placed in different parts of the circumference. Thus the requiſite heat may be maintained, and the chickens uſhered into life.

If we would be œconomical of heat, and draw utility from what is commonly loſt, we may employ, for artificial incubation, that of the ovens for bakers and paſtry-cooks, of forges, and glaſs-houſes, and even that of a chamber-ſtove, or a kitchen-grate, conſtantly keeping in mind that the ſucceſs will depend chiefly on the equal diſtribution of heat, and the total excluſion of humidity.

When the ovens are large and well-managed, they exclude thouſands of chickens at a time. This profuſion would be rather inconvenient in a climate like ours, if we had not as well diſcovered a method of rearing the brood independently of the aſſiſtance of the hen, as of hatching them without her incubation. This conſiſts in a more or leſs perfect imitation of the manner

ner in which the mother treats her young after they have emerged from the ſhell.

We ſhall eaſily conceive, that the mother which ſhews ſo ſtrong an inclination to cover her eggs, ſits on them with ſuch unremitting affiduity, and takes ſo lively a concern for embryos that have yet no being, will not cool in her attachment after her chickens are hatched. Her affection is heightened by the ſight of theſe little creatures, that owe to her their exiſtence; it is every day increaſed by the repetition of cares which their feeblenefs requires. Continually employed in watching over them, ſhe ſeeks food, merely to ſatisfy their craving wants. If ſhe cannot find it, ſhe ſcrapes the earth with her nails to extract the nouriſhment concealed in its boſom, and freely beſtows it on her young. She recals them when they wander, ſpreads her wings over them to defend them againſt the inclemency of the weather, and broods a ſecond time. She enters into theſe tender concerns with ſo much ardour and anxiety, that her health is viſibly impaired, and ſhe can be diſtinguiſhed from every other hen by her ruffled feathers, her trailing wings: and the hoarſeneſs of her voice, and the different inflections, are all expreſſive of her ſituation, and mark ſolicitude and maternal affection.

But if ſhe neglects herſelf in preſerving her young, ſhe expoſes herſelf to every danger in their defence. If a ſparrow-hawk appear in the air this mother, fo feeble, fo timid, and which in every other circumftance would confult her fafety by flight, becomes intrepid, from the warmth of attachment, darts before the dreaded talon, and by her repeated cries, the clapping her wings, and her undaunted refolution, fhe often intimidates the rapacious bird, which, re-pulfed by the unexpected refiftance, retires to feek eafier prey. She then feems to poffefs all the qualities of a generous mind; but what re-flects no lefs honour on her inftinct is, that if fhe has been made to fit on ducks eggs, or thofe of any other water-foul, her affection is no lefs ardent for thefe ftrangers than for her own progeny. She does not perceive that fhe is only their nurfe, and not their mother; and when, directed by nature, they plunge into the neighbouring ftream, it is amufing to obferve the aftonifhment, uneafinefs, and vexation of the poor nurfe, who fancies fhe is ftill a mo-ther; impelled on the one hand by the defire of following them into the midft of the water, checked on the other by the invincible repug-nance to that element, teafing herfelf with fruitlefs buftling along the margin, trembling, forlorn, beholding her family in imminent dan-ger, and not daring to afford them help.

It would be impoffible to fupply all the affi-duities of the hen in raifing her young, if the fame degree of attention were required, and an equal affection with that of the mother. We
fhall

fhall fucceed by noticing the principal circumftances in the conduct of the hen to her progeny, and by imitating them as much as we can.
For example, it being obferved, that the chief
object of the mother's attention is to lead the
chickens where they can find food, and to guard
them againft cold and the inclemency of the
weather; a plan has been formed to effect this,
and with even more advantage than from the
affiftance of the hen herfelf. If they be hatched
in winter, they are kept a month or fix weeks
in a ftove, heated to the fame degree with the
ovens for incubation, and only let out five or fix
times a-day to eat in open air, and efpecially to
enjoy the fun. The warmth of the ftove pro-
motes their growth, and the expofure to the air
ftrengthens and braces them. Crumbs of bread,
yolks of eggs, and millet feeds, are their firft
food. If the incubation be performed in fummer, they are kept in the ftove only three or four
days; and in all feafons they are brought out
of the ftove only to be put into the crib, which
is a kind of fquare cage, fhut with a front of
grating made of iron wire, or fimple netting, and
clofed above with a hinged lid. In this cage
the chickens are fed; but after they have eaten
enough, and taken fufficient exercife, they muft
be fheltered fo as to allow them to enjoy warmth
and repofe. Hence the chickens that are led by
their mother are accuftomed to affemble under
the covert of her wings. For this purpofe Rea-

mur

mur contrived an *artificial mother*; this is a box
formed of fheeps fkin, the bottom of which is
fquare, and the upper part floped like the top of
a defk.   He places this box on one of the ends
of the crib, in fuch a fituation that the chickens
may enter eafily, and walk round the three fides
at leaft; he warms it below by means of a foot-
ftove, which he refrefhes occafionally.   The inclination of the cover of this kind of defk allows
the chickens to place themfelves according to
their different fizes; but as they have a practice,
efpecially when they are cold, of preffing together, and even climbing on each other, and in
this crowd the weak and fmall ones run the rifk
of being fmothered, this *artificial mother* is kept
open at both ends, or rather both ends are covered with a net which the leaft chicken can
remove, fo as eafily to efcape when it feels itfelf
too much fqueezed; and it can then, by going
round to the other hole, chufe a lefs dangerous
place.   Reamur endeavours to avoid even this
inconvenience by another precaution, which is
to keep the cover floped fo low as to prevent the
chickens from climbing on each other; and he
raifes it gradually as they grow.   He improves
ftill on this idea, by dividing his largeft cribs
into two, by means of a tranfverfe partition, fo
as to be able to feparate the chickens of different
fizes.   He even places the cribs on wheels, that
they may be eafily tranfported; for they muft
be brought into a chamber always in the evenings,

ings, and even during the day-time when the
weather is bad; and in winter this chamber
muft be warmed. But when it is neither cold
nor rainy, the cribs ought to be expofed to the
open air and the fun, only taking the precaution of fheltering them from the wind. The
doors may even be left open, and the chickens
will foon learn to come out to fcrape the dung,
or peck the tender grafs, and return to their
food, or to recover their warmth under their
artificial mother. If we would not hazard their
fauntering at will, we may place at the end of
their crib an ordinary hen-crib, which, com-
municating with the firft, will allow them a confiderable fpace to roam in, and at the fame time
prevent accidents.

But the more we confine them, the more regular we ought to be in giving them food. Befides millet feeds, yolks of eggs, foup, and crumbs
of bread, young poultry are fond of rape-feed,
hemp-feed, and other fmall grains of that kind;
peafe, beans, lentils, rice, fhelled barley and oats,
chopped turkey beans, and buck-wheat. It is
proper, and even œconomical, to burft moft of
thefe in boiling water, before we offer them;
the faving amounts to a fifth on the wheat, two
fifths on the barley, one half on the turkey beans,
but is nothing on the oats and buck-wheat. It
would even be a lofs to foak the barley; but this
grain is what the chickens fhew the moft indifference for. Laftly, after they have grown, we

may

may give them every thing that we eat ourfelves,
except bitter almonds * and coffee-beans †. Every kind of chopped flefh, whether raw or boiled,
but particularly earth-worms, are proper; they
difcover fo great a liking for this fort of food,
that one would imagine that they are carnivorous, and perhaps nothing is wanting to them,
as well as to many others, but a hooked bill
and claws, to conftitute them real birds of prey.

It muft however be admitted, that poultry
differ from birds of prey, no lefs by their mode
of digeftion and the ftructure of their ftomach,
than by their bill and their nails.   In thefe the
ftomach is membranous, and digeftion is effected
by means of a folvent, which varies in different
fpecies, but the action of which is well afcertained ‡ ; whereas the gallinaceous tribe may
be confidered as having three ftomachs; *viz.*
1. The craw, which is a kind of membranous
bag, where the grains are firft macerated and begin to be reduced to a pap : 2. The wideft part
of the canal, lying between the craw and the
gizzard, but neareft the laft; it is lined with a

* Ephemerides des Curieux de la Nature, Dec. 1. an. 8. obf. 99.
† " Two chickens being fed, the one on burnt coffee from the
iflands, the other on frefh coffee, they both grew confumptive and
died, the one on the eighth day, and the other on the tenth, after
having eaten three ounces of coffee. Their feet and legs were
fwelled, and the gall-bladder as large as that of a turkey cock."
Memoires de l'Academie Royale des Sciences, année 1746, p. 101.
‡ Memoires de l'Academie Royale des Sciences, an. 1752,
p. 266.

number

number of fmall glands, which furnifh a liquor
that the food imbibes in its paffage : 3. Laftly,
The gizzard, which yields a liquor that is manifeftly acid, fince the internal coat, being foaked
in water, becomes an excellent runnet for curd-
ling milk. This third ftomach completes, by the
powerful action of its mufcles, what had only been
begun in the two firft. The force of its fibres is
greater than could be conceived; in lefs than
four hours a ball of glafs, which could fuftain
a preffure of four pounds, is reduced to an im-
palpable powder. In forty-eight hours, feveral
tubes of glafs, four lines in diameter and one
line thick, were divided longitudinally into two
kinds of rents ; and, at the end of that time, all
the fharp edges were ground down, and the polifh deftroyed, particularly on the convex part.
The gizzard was alfo able to flatten tubes of
tinned iron, and, in the fpace of twenty-four
hours, to crufh feventeen nuts ; and this was
effected by repeated compreffions and alternate
attrition, the mechanifm of which it is difficult
to perceive. Reaumur, who made feveral trials
to difcover it, never could diftinguifh but once
any confiderable motions in that part. He
faw in a capon the gizzard, of which he had
brought into view portions, contracting and fink-
ing, and again fwelling ; he obferved a kind of
flefhy chords which formed on the furface, or rather appeared to be forming, becaufe he made in-
cifions between them which feparated them ; and

I   all

all thefe motions appeared to be propagated in
waves, and very flowly.

What proves that, in gallinaceous birds, digeftion is performed chiefly by the action of the
mufcles of the gizzard, and not by the power of
any folvent, is, that if one of thefe be made to
fwallow a fmall lead tube, open at both ends, but
fo thick as to refift the compreffion of the giz-
gard, and into which a grain of barley be introduced, the tube will be found in the fpace of two
days to have loft confiderably of its weight; but
the grain inclofed, though it were boiled and
fhelled, will then be difcovered to be fomewhat
fwelled, but as little altered as if it had been left
the fame time in another place equally humid;
whereas the fame grain, and others that are much
harder, if not protected by a tube, would be di-
gèfted in much lefs time.

One circumftance which may affift the action
of the gizzard is, that birds keep the cavity as
full as poffible, and thus the four mufcles of
which it confifts are thrown into play. When
grain is wanting, they cramb it with herbage,
and even fmall flints, the hardnefs and roughnefs of which contribute to bruife the grain
againft which they are inceffantly rubbed. I fay
by their roughnefs; for, when they are polifhed, they quickly pafs through the body, and
thofe with rugged furfaces only remain. They
are the more numerous in the gizzard the fcarcer
the food is, and they continue in it a longer
time

time than any other fubftance, whether digeft-
ible or indigeftible.

We fhall not be furprifed that the inner coat
of this ftomach is ftrong enough to refift the re-
action of fo many hard bodies on which it conftantly grinds, if we confider that it is really very
thick, and of a fubftance analogous to horn. Befides, we know that bits of wood and leather,
which are rubbed with an extremely hard powder to polifh bodies, laft for a very long time.

We may alfo fuppofe that this membrane is
renewed in the fame manner as the callous fkin
of labourers hands.

But though the fmall ftones may affift digeftion, it is not afferted that the granivorous
birds have a decided view in fwallowing them.
Redi having fhut up two capons, with water and
little pebbles for their food; they drank much
water, and died; the one in twenty days, the
other in twenty-four, and both of them without having fwallowed a fingle ftone. Redi found
many in the gizzard, but thefe were what had
been fwallowed before.

The organs that are deftined for refpiration,
confift of lungs, fimilar to thofe of the land animals, with ten air cells, eight of which are within the breaft, and communicate directly with the
lungs; and two larger ones in the lower belly,
and which communicate with the eight preceding. When in infpiration the thorax is dilated,
the air enters by the larynx into the lungs, thence

into

into the eight upper air cells, which dilating in-
hale that alſo of the two cells of the lower belly,
and theſe ſuffer a proportional collapſe.   When,
on the contrary, the lungs and upper cells, contracting during expiration, preſs the air included
in their cavity, it eſcapes partly through the
larynx, and partly returns from the eight cells
in the breaſt into the two in the lower belly,
which then dilate by a mechaniſm nearly analogous to that of a double bellows.   But this is
not the proper place to explain the mechaniſm;
it will be ſufficient to obſerve, that in thoſe birds
which never fly, as the caſſowary, the oſtrich,
&c. and in thoſe that fly tardily, ſuch as the
gallinaceous tribe, the fourth cell on each ſide is
the ſmalleſt *.

All theſe differences in the ſtructure neceſſari-
ly imply many others, not to ſpeak of the membranous tubes that are obſerved in ſome birds.
Duverney has ſhewn, from an experiment made
on a living cock, that the voice in theſe birds is
formed not near the larynx, as in the quadrupeds, but below the *trachea arteria*, near the
forking, at which place Perrault perceived an internal larynx.   Heriſſant obſerved, in the principal bronchial veſſels of the lungs, ſemilunar
membranes placed tranſverſely one above another, in ſuch a manner that they only occupied
the half of the cavity of theſe veſſels, and allow-

* Memoires pour ſervir a l'Hiſtoire les Animaux.

ed

ed the air a free paffage through the other half; and he juftly concludes, that thefe membranes muft contribute to the formation of the voice, though their affiftance is not fo effential as that of the offeous coat of the crefcent, which terminates a confiderable cavity above the fuperior and internal part of the breaft, and which has alfo fome communication with the upper air cells. This anatomift affirms, that he has afcertained, by repeated trials, that if this coat be perforated, the voice is immediately impaired, and can be reftored only by clofing the hole accurately to vent the efcape of the air *.

After obferving fuch wide differences in the conformation of the organs of the voice, will it not appear fingular, that birds, with a tongue of cartilage, and lips of horn, fhould imitate our fongs, and even our fpeech, more eafily than thofe among the quadrupeds that refemble man the moft? So difficult it is to judge of the ufe of the parts from their mere ftructure, and fo true, that the modification of the voice and of founds depends almoft entirely on the fenfibility of the ear.

The inteftinal canal is very long in the gallinaceous tribe, and exceeds about five times that of the animal, reckoning from the point of the bill to the *anus*. We find two *cæca*, about fix inches in length, which take their rife where the

---

* Memoires de l'Academie Royale des Sciences, *année* 1753, page 191.

colon

colon joins the *ileon;* the *rectum* widens towards
its extremity, and forms a common receptacle,
into which the folid and fluid excrements are
feparately difcharged, and from which they are
ejected together, though not intimately mixed.
The organs characteriftic of the fexes are alfo
perceived ; viz. in hens, the *vulva* or orifice of
the *oviductus,* and in Cocks the two yards, that
is, the paps of the two fpermatic veffels. The
vulva, as we have before mentioned, is placed
above the *anus,* and confequently the difpofition
of thefe parts which obtains in quadrupeds is reverfed.

It was known in the time of Ariftotle, that
the Cock had tefticles concealed within its body.
The ancients even afcribed to this fituation the
fiery paffion of the male for the female, who is
lefs ardent, they alleged, becaufe the ovarium being placed near the diaphragm, is more apt to
be cooled by the acceffion of the air refpired *.
But the tefticles are not fo exclufively appropriated to the male as not to be found in the females of fome fpecies of birds, as in the little
buftard, and perhaps in the great buftard †.
Sometimes the male has only one, but generally
two ; the bulk of thefe kinds of glands is far
from being proportioned to that of the bird. In
the eagle, they are only of the fize of peas ; in

* Ariftotle *de Partibus Animalium,* lib. iv. 5.
† Hiftoire de l'Academie Royale des Sciences, *année* 1756,
p. 44.

a Cock

a Cock of four months old, they are as large as
olives. The bulk varies not only between one
fpecies and another, but between different individuals of the fame fpecies, and is moft expanded
in the feafon of love. But how diminutive foever
be their fize, they produce mighty effects in the
animal œconomy, as is evinced by the wonderful changes effected by their extirpation. This
operation is commonly performed when the bird
is three or four months old. After emafculation
it grows plumper, and its flefh becomes more
juicy and delicate; and when fubjected to a che-
mical analyfis, yields different products from
what it would have given before caftration*.
The capon is no longer liable to moult; in the
fame manner as the buck, when degraded from
his fex, never cafts his antlers. The note is altered, his voice is broken, and feldom heard:
treated roughly by the Cocks, with difdain
by the females, deprived of all the appetites related to generation, it is not only excluded from the fociety of its equals, but extruded, as it were, from its fpecies. It is an idle
folitary out-caft, all whofe powers are directed
on itfelf, and whofe fole object is its individual

* The extract of the lean of a capon is fomewhat lefs than the
fortieth part of its total weight; whereas it amounts to one twelfth
in a pullet, and rather more than one feventh in a Cock. Befides,
the extract of Cock's flefh is very dry, while it is difficult to feparate the humidity from that of a capon.
Memoires de l'Acad. Royale des Sciences, année 1730, p. 231.

preferva-

prefervation : to eat, to fleep, to fatten, are its
principal employments, and conftitute the fum
of its wants. But, by a little attention, we can
draw advantage from its weaknefs, and even its
confequent docility, by giving it ufeful habits.
For inftance, we can teach it to rear and tend
young chickens. For this purpofe it muft be
kept fome days in a dark place, only bringing
it out at regular hours to feed, and accuftoming
it gradually to the fight and company of a few
chickens that are pretty ftout ; it will foon contract a fondnefs for them, and will lead them with
as much affection and affiduity as their mother.
It will even conduct a greater number than a
hen ; for its wings fpread and afford more fhelter ; and the hen, freed from its toil and folicitude, will foon begin again to lay ; and thus
the capon, though condemned to fterility, will
ftill contribute indirectly to the prefervation and
multiplication of its fpecies.

So great a change produced in the character
of the capon, by a caufe fo weak and apparently
fo inadequate to the effect, is the more remarkable, as it is confirmed by an immenfe number
of trials which men have made on other fpecies,
and have even dared to extend to their brethren
of the human race.

The Cock has been the fubject of another experiment that is far lefs cruel, but perhaps no
lefs important for the fcience of phyfiology : it
is,

is, after cutting the comb * as ufual, to fubftitute
in its ftead one of the young fpurs which has
juft begun to fhoot; thus engrafted, it gradually
ftrikes root into the flefh, thence extracts its
nourifhment, and often grows more luxuriantly
than it would in its natural place. Some have
attained to two inches and a half in length, and
more than three lines and a half in diameter at
the bafe; fometimes they are twifted round like
the horns of a ram, at other times bent backwards like thofe of a he-goat †.

This is a kind of animal engrafting, the fuccefs of which would appear very doubtful when
firft tried, but from which, fince the fuccefs is
known, it is aftonifhing that no practical information has been derived. In general, the deftructive experiments have been more ftudied,
and purfued with more ardour, than thofe which
are directed to prefervation; becaufe man is
fonder of pleafure and expence, than the acqui-
fition of knowledge and the exercife of bene-
ficence. Chickens are not hatched with that
creft and thofe reddifh membranes which diftinguifh them from other birds. It is a month
after they have left the fhell before thefe parts
begin to unfold; at two months old, the young

* The reafon why the capon's comb is cut is, that, after emafculation, it does not ftand erect but becomes pendulous, and would
therefore prove inconvenient by hiding an eye.

† Arciens Memoires de l'Academie Royale des Sciences,
tom. xi. p. 48. Journal Economique, Mars 1761, p. 120.

Cocks

Cocks crow, and fight with one another. They feel that they muſt hate each other, though the ſource of their diſlike has yet no exiſtence. It is commonly five or ſix months before they ſhew any paſſion for the hens, and that theſe begin to lay. In both ſexes, the complete term of their growth is a year, or fifteen months. The young hens, it is ſaid, lay more; but the old ones are better ſitters. This period of their growth would imply that the ordinary extent of their life does not exceed ſeven or eight years, if the ſame proportion ſubſiſted in birds as in quadrupeds. But we have ſeen that this is much longer; a Cock will live twenty years in the domeſtic condition, and perhaps thirty years in the ſtate of liberty. Unfortunately for them, we have no intereſt in ſuffering them to reach to a great age. The hens and capons that are deſtined for the table, never enjoy above one year's exiſtence; and moſt of them have only one ſeaſon. Thoſe which are ſelected for the multiplication of the ſpecies, become ſoon exhauſted, and none are permitted to finiſh the period aſſigned by nature; ſo that it is a ſingular accident, that cocks have ever been ſeen to die of age.

Poultry can ſubſiſt in all places under the protection of man, and accordingly they are ſpread over the whole inhabited world. The better ſort of people breed hens in Iceland *,

* Horrebow's Deſcription of Iceland.

where

where they lay as in other parts; and the warm
countries abound with them. But, according to
Dr. Thomas Hyde *, Perfia is the native region
of Cocks; there they are numerous, and held
in great eftimation, efpecially by certain dervifes, who confider them as living clocks; and
it is well known that a clock is the foul of
every fociety of dervifes.

Dampier fays, that he faw and killed, in the
iflands of Poulocondor, wild Cocks, that were
not larger than our crows; and whofe crow was
much like that of our dunghill Cocks, only
fhriller. In another part he adds, that there are
fome in the ifland of Timor, and at St. Jago,
one of the Cape de Verd iflands. Gemelli
Carreri relates that he obferved fome in the
Philippines; and Merolla afferts that there are
wild hens in the kingdom of Congo, which are
more beautiful, and have a more delicate flavour, than our domeftic kind; but that the
Negroes fet little value on thefe birds.

From their native climate, wherefoever it be,
thefe birds have fpread over the extent of the
ancient Continent, from China to Cape Verd;
and from the Southern Ocean to the Seas of the
North. Thefe migrations were performed in
remote ages, far beyond the reach of hiftorical

* *Hiftoria Religionis veterum Perfarum.* Obferve, however,
that the art of fattening capons was introduced by the Armenian
merchants into Perfia from Europe, as appears from Tavernier.

tradi-

tradition. But their fettlement in the New
World feems to be a much more recent event.
The hiftorian of the Incas informs us, that there
were none at Peru before its conqueft, and that
after a refidence of more than thirty years, the
hens could not be habituated to hatch in the
valley of Cufco. Coreal pofitively afferts, that
poultry were introduced by the Spaniards into
Brazil, and that the inhabitants of that country
would eat none of them, and looked upon their
eggs as a kind of poifon. Nor, according to
the teftimony of F. Charlevoix, had the natives
of St. Domingo any ; and Oviedo confiders it
as a certain fact, that they were carried from
Europe to America. Acofta indeed maintains
the oppofite opinion, that hens exifted in Peru
before the arrival of the Spaniards ; and alleges
as a proof, that the natives call the bird, in their
language, *gualpa*, and its egg *ponto*. But the
antiquity of the word is not fufficient to eftablifh
that of the thing denoted ; for it is eafy to conceive, that favages, the firft time they faw a
ftrange bird, would naturally give it a name,
either from its refemblance to fome bird with
which they were acquainted, or from fome
other analogy. What would determine me in
favour of the firft opinion is, the conformity to
the law of the climate. This law, though it
cannot be applied in general to birds, efpecially
thofe which are vigorous on the wing, and to
which all countries are open, yet regulates thofe
which,

which, like the poultry, being bulky and having an averſion to water, can neither waft their
courſe through the air like the birds that ſoar,
nor croſs the ſeas, nor even the large rivers,
like the quadrupeds that ſwim, and would
therefore be for ever excluded, but for the inter-
ference of man, from thoſe countries which are
ſeparated by an immenſe ocean. The Cock is
then an animal which belongs peculiarly to the
Ancient Continent, and ought to be added to
the liſt that I have given, of all thoſe animals
which exiſted not in the New World before it
was diſcovered.

In proportion as hens are removed from their
native region, and accuſtomed to another climate and different food, they muſt ſuffer ſome
alteration in their ſhape, or rather in the parts
moſt ſuſceptible of change. Hence undoubtedly thoſe varieties that conſtitute the different
breeds which I am to deſcribe ; varieties which
are conſtantly perpetuated in each climate, whether from the continued action of the ſame
cauſes that produced them at firſt, or from the
attention that is paid in matching the individuals
ſelected for propagation.

It is to be wiſhed that we could here form, as
in the caſe of the dog, a kind of genealogical
tree of all families of the Cock, which would
point out the primitive ſtock, and its different
branches, and repreſent the various orders of
alterations and changes correſponding to its dif-

ferent

ferent ftates. But this would require more accurate and more minute accounts than thofe in
our poffeffion. I fhall therefore content myfelf
with giving my opinion in regard to the hen of
our own climate, and fhall endeavour to examine into it's origin; but previous to this
inquiry, I fhall enumerate the foreign breeds that
have been defcribed by naturalifts, or only
mentioned by travellers.

1. *The Common Cock.* That of our own
climate.

2. *The Crefted Cock.* It is diftinguifhed from
the Common Cock by a tuft of feathers rifing
on the head, and by its comb, which is generally fmaller; probably becaufe the food, inftead of being fpent on the comb alone, is partly
diftributed to nourifh the feathers. Some travellers affert that the Mexican poultry are
crefted; thefe, as well as all the reft on the continent of America, have been introduced from
the ancient continent. The breed of the crefted
hens is that which the curious have moft cultivated, and what generally happens when things
are clofely examined, they have obferved a great
number of differences, particularly in the colours of their plumage; which ferve to diftinguifh a multitude of races, that are the more
efteemed in proportion to the beauty and rare-
nefs of their tints. Such are the gold and filver
ones; the black-crefted white ones; the white-
crefted black ones; the agate, the chamois, and
the

the flaty; thofe with fifh-fcales, and the er-
mined; the widow-hen, which has fmall white
tears fprinkled on a fallow ground; the flame-
coloured; the ftony fort, whofe plumage has a
white ground fpotted with black, chamois,
flaty, or golden, &c. But I fufpect that thefe
differences are not fo invariable, or fo deeply
impreffed, as to conftitute real diftinct fpecies,
as fome curious people pretend, who affert that
many of the above breeds never intermix.

3. *The Wild Cock of Afia.* This is undoubtedly what approaches the neareft to the original
ftock of our common kind; for never being fet-
tered by man, or thwarted in the choice of its food
or mode of life, what could ever alter its native
purity? It is neither the largeft nor the fmalleft
of its fpecies, but is intermediate between the
extremes. It is found, as we have already
obferved, in many countries of Afia, in Africa,
and in the Cape de Verd iflands. We have no
defcription of it fo complete as to enable us to
compare it with our Cock. I muft here recom-
mend to travellers who have an opportunity of
feeing this wild breed, to inquire if they conftruct nefts, and in what manner. Lottinger,
phyfician of Strafburg, who has made many
important obfervations on birds, informs me,
that our hens, when left to themfelves, build
nefts, and with as much care as the partridges.

4. *The*

4. *The Acoho,* or *Madagafcar Cock.* This fpecies is very fmall, and the eggs ftill lefs in proportion, for the birds can hatch thirty at a time *.

5. *The Dwarf Hen of Java.* It is of the fize of a pigeon †; probably the Little Englifh Hen, which is ftill fmaller than the Dwarf Hen of France, is of the fame kind. We may, perhaps, add the Small Hen of Pegu, which travellers defcribe as not larger than a middle-fized turtle; its feet fcaly, and its plumage beautiful.

6. *The Hen of the Ifthmus of Darien.* It is fmaller than the common fort; has a circle of feathers round its legs, an exceeding thick tail, which it carries erect, and it crows before break-of-day.

7. *Cambogia Hens.* Carried by the Spaniards from that kingdom into the Philippines. Their feet are fo fhort that their wings trail on the ground. It is very like the Dwarf Hen of France, or perhaps that Dwarf Hen that is reared in Britanny, on account of its fecundity, and which conftantly hops in its gait.

8. *The Bantam Cock.* It has much refemblance to the Rough-footed Cock of France. Its feet are covered with feathers, but only on the

* Hiftoire Generale des Voyages, tom. viii.
† Collection Academique, partie etrangere, tom. iii.

outfide.

outfide. The plumage of the legs is very long, and forms a fort of boots which reach a confiderable way beyond the claws. It is courageous, and refolutely fights with one ftronger than itfelf. Its iris is red. I have been informed that moft of this breed have no tuft. There is a large kind of rough-footed Cocks that comes from England, and another fmaller, termed the *Englifh Dwarf Cock;* which is of a fine gold colour, with a double comb.

There is ftill another fort of dwarfs, which exceeds not the fize of a common pigeon, 'and whofe plumage is fometimes white, fometimes mixed with white and gold colour.

9. The Dutch fpeak of another kind of Cocks peculiar to the ifland of Java, where they are feldom reared but for fighting; they call it the *Half-Hen of Java.* According to Willoughby, it carries its tail nearly like the turkey. To this family we muft refer thofe fingular hens of Java, mentioned by Mandeflo, which refemble the common and Indian kinds, and that fight defperately with each other like cocks. The Sieur Fournier informs me, that one of this fpecies is ftill living at Paris; it has, according to him, neither comb nor ruff; the head is fimple like that of the pheafant. This hen is very high on its legs; its tail is long and pointed, and the feathers of unequal length; and in general, the colour of the feathers is auburn, like thofe of the vulture.

10. *The*

10. *The English Cock* is not larger than the Dwarf Cock, but is much taller than our Common Cock, and this is what principally diftinguifhes it. We may alfo clafs with it the *Xolo* *, a kind of Philippine Cock, which has very long legs. Befides the Englifh Cock excels the French in fighting; it has rather a tuft than a creft; its neck and bill are more flender; and above the noftrils there are two flefhy protuberances, which are red like the comb.

11. *The Turkish Cock* is remarkable only for its fine plumage.

12. *The Hamburgh Cock*, named alfo *the Velvet Breeches*, becaufe its thighs and belly are of a foft black. Its demeanour is grave and ftately; its iris is yellow, and its eyes are encircled with a ring of brown feathers, from which rifes a black tuft that covers the ears. There are other feathers nearly like thefe behind the comb and beneath the barbils, and broad round black fpots on the breaft. His legs and feet are of a lead colour, excepting the fole of the foot, which is yellowifh.

13. *The Frizzled Cock*, whofe feathers have a reverfed pofition. They are found in Java, Japan, and the whole of the fouth of Afia. This bird belongs more peculiarly to the warm countries; for chickens of this breed are extremely fenfible to cold, and can hardly fupport that

* Gemelli Carreri.

of

of our climate. The Sieur Fournier assures me, that their plumage assumes all sorts of colours, white, black, silvery, golden, and slate tints.

14. *The Silky Hen of Japan.* The feathers are white, and their webs are parted, and pretty much resemble hair. Its feet are clothed with plumage on the outside, as far as the nail of the outer toe. This breed is found in Japan, China, and in some other countries of Asia. To propagate it in all its purity, requires that both the parents be covered with down.

15. *The Negro Cock* has its comb, barbils, epidermis, and periosteum absolutely black. Its plumage also is generally black, but sometimes white. It is found in the Philippines, in Java, Delhi, and at St. Jago, one of the Cape de Verd islands. Becman affirms that most of the birds in the last mentioned place have bones as black as jet, and a skin black like that of a negroe*. If this fact be true, we must impute it to the tinging quality of their food. We know the effects of madder, and other plants of that genus, and we are informed, that in England the veal is whitened by feeding the calves with grain and other soft aliments, mixed with a certain earth or chalk found in the county of Bedford †. It would therefore be curious to discover at St. Jago, among the different substances which these birds eat, that which tinges the *periosteum* black. This negroe hen is also known in France, and pro-

* Dampier.　　† Journal Economique, *Mai* 1754.

pagates

pagates there; but as its flefh, after being dreffed, is black and unpalatable, it is not likely that the race will be multiplied. When it croffes the breed with others, varieties of different colours are produced, but which commonly retain the comb and the ruffs, or black barbils; and even the membrane that forms the little ear is tinged with a blackifh hue on the outfide.

16. *The Rumplefs Cock*, or *the Perfian Cock* of fome authors. Moft of the hens and Cocks in Virginia have no rump, and yet they are un-doubtedly of the Englifh breed. The inhabi-tants of that colony affirm, that when thefe birds are imported, they foon lofe the rump *. If this be admitted as a fact, the variety in queftion ought to be called *Virginian*, and not *Perfian Cocks;* efpecially as they were unknown to the ancients, and the moderns have not noticed them till after the difcovery of America. We have mentioned that the European dogs, which have pendulous ears, lofe their voice, and that their ears become erect, when they are carried into tropical climates. This fingular change, pro-duced by the exceffive heat of thofe torrid re-gions, is not however fo great as the lofs of the rump and tail in the gallinaceous tribe. But it appears to me much more curious, that as thefe two tribes of animals are the moft domeftic of all, and therefore the wideft removed from their

---

* Philofophical Tranfactions for 1693, No. 206.

natural

natural condition, fo there is breed of dogs without a tail, as there is of Cocks without a rump.
Several years ago, I was fhewn one of thefe dogs
that had been whelped without a tail, but which I
then conceived to be a degraded individual, a
monfter; and for that reafon I took no notice
of it in the hiftory of the canine genus. I have
again confidered the fubject, and I am now confident that it is a conftant and invariable race,
like that of the Rumplefs Cock. This breed of
Cocks has a blue bill and feet; a fingle or double
comb, but no creft; the plumage is variegated
with all the colours; and the Sieur Fournier
affures me, that when it couples with the ordinary
kind, a half-rumped fort are produced, with fix
feathers in the tail inftead of twelve.—This may
be true, but I can hardly believe it.

17. *The Hen with five toes* is, as we have faid,
a powerful objection to the fyftem of claffifica-
tion founded on the number of toes. This kind
has five on each foot, three before and two behind; there are even fome individuals that have
fix.

18. *The Hens of Sanfevara.* The eggs of this
fort are fold in Perfia for three or four crowns
a-piece; and the Perfians amufe themfelves in
ftriking them againft one another, as a kind of
play. In that country there are alfo Cocks much
more beautiful and larger, which are fold fo high
as 300 crowns *.

* Tavernier.

19. *The*

19. *The Cock of Caux*, or of *Padua*. Its diſtin-guiſhing character is its magnitude. Its comb is often double, and in the ſhape of a crown; and there is a kind of tuft, which is more conſpicu-ous in the hens. Their voice is ſtrong, hollow, raucous, and their weight is from eight to ten pounds. To this fine breed we may refer the large Cocks of Rhodes, Perſia *, Pegu †, the bulky hens of Bahia, which do not begin to be covered with feathers till they have attained half their ſize ‡. It is well known that the hens of Caux are not ſo ſoon feathered as the ordinary ſort.

It may be obſerved, that a great number of birds, mentioned by travellers by the names of Cock and Hen, are of a quite different ſpecies. Such are the *patonardes* or *palonardes* found near the Great Bank, and which are ſo fond of cods liver ‖; the Cock and hen of Muſcovy, which are the male and female grous; the red hen of Peru, which is like the pheaſant; that large tuft-ed hen of New Guinea, whoſe plumage is of an azure blue, which has a pigeon's bill and feet like thoſe of the common poultry, and which neſtles in trees §, and is probably the Banda phea-

* Chardin.

† Recueil des Voyages qui ont ſervi à l'Etabliſſement de la Com-pagnie des Indes, *tome* iii. p. 71.

‡ Dampier's New Voyage.

‖ Recueil des Voyages du Nord, *tome* iii. p. 15.

§ Hiſtoire generale des Voyages, *tome* xi. p. 230.

ſant;

fant; the hen of Damietta, which has a red bill
and feet, a fmall fpot on the head of the fame
colour, and plumage of a violet blue, and which
muft be confidered as a great water-fowl; the hen
of the Delta, the rich colours of whofe plumage
Thevenot extols, but which differs from the
common fort, not only by the fhape of its bill
and tail, but by its natural habits, fince it is fond
of marfhes; the Pharaoh hen, which the fame
traveller affirms is not inferior to a fat hazel
grous; the hens of Corea, which have a tail
three feet long, &c.

Amidft the immenfe number of different breeds
of the gallinaceous tribe, how fhall we determine the original ftock? So many circumftances
have operated, fo many accidents have con-
curred; the attention, and even the whim of
man have fo much multiplied the varieties, that
it appears extremely difficult to trace them to
their fource. The Wild Cocks found in the warm
countries of Afia may indeed be confidered as
the primeval ftem in thofe regions. But as in
our temperate climates there is no wild bird that
perfectly refembles the Domeftic Cock, we are
at a lofs on which of the varieties to confer
the priority. The pheafant, the grous, the wood-
hen, are the only birds in the ftate of nature
which are analogous to our poultry; but it
is uncertain if they would ever intermix, and
have prolific progeny; and they have confti-
tuted diftinct and feparate fpecies from the moft

remote

remote times. Befides, they want the combs, the fpurs, and the pendulous membranes of the gallinaceous tribe.   If we exclude all the foreign and wild kinds, we fhall greatly diminifh the number of varieties, and the differences will be found to be flight.   The hens of Caux are almoft double the bulk of the ordinary fort ; the Englifh Cock, though exactly like the French, has much longer legs and feet ; others differ only in the length of their feathers ; others in the number of their toes ; others are diftinguifhed by the beauty and fingularity of their colours, as the Turkifh and Hamburgh hen : and of thefe fix varieties, to which the common breed may be reduced, three are to be afcribed to the influence of the climate ; that of Hamburgh, that of Turkey, and that of England ; perhaps alfo the fourth and fifth, for the hen of Caux moft probably came from Italy, fince it is alfo called the *hen of Padua*, and the hen with five toes was known in Italy in the time of Columella. Thus there only remain the Common Cock and the Crefted Cock as the natural breed of our country, and even in thefe the two fexes admit of all the variety of colour.   The conftant character of the tuft feems to mark an improved fpecies ; that is, one better kept and better fed ; and confequently the common breed, which has no tuft, muft be the true parent of our poultry. It would appear that the primitive colour was white, and that all the intermediate fhades between

tween it and black were fucceffively affumed.
What feems to corroborate this conjecture is, an
analogy which no perfon has yet remarked, that
the colour of the egg generally refembles that of
the plumage of the bird. Thus a raven's eggs are
of a green brown, fpotted with black; thofe of
the keftril are red; thofe of the caffowary
dark green; thofe of the black crow are of a
ftill duller brown than thofe of the raven; thofe
of the variegated magpie are alfo variegated and
fpotted; thofe of the great cinereous fhrike, fpotted with grey; thofe of the woodchat, fpotted
with red; thofe of the goatfucker, mottled with
bluifh and brown fpots on a cloudy whitifh
ground; thofe of the fparrow, cinereous entirely, covered with chefnut fpots on a grey ground;
thofe of the blackbird, blackifh blue; thofe of
the grous, whitifh fpotted with yellow; thofe of
the pintados, fpeckled like their plumage, with
white round fpots, &c. In fhort, there feems
to be an invariable relation fubfifting between
the colour of the egg and that of the plumage.
The tints are indeed much more dilute on the
eggs, and in moft of them the white predomi-
nates; but white is alfo in moft cafes the prevailing colour of the plumage, efpecially in females: and fince hens of all colours, white, black,
grey, tawney, and mottled, have white eggs, there
is reafon to conclude, that if they had remained
in the ftate of nature, white would at leaft have
predominated in their plumage. Domeftication

has introduced various fhades on the feathers;
but as thefe are only accidental and fuperficial,
they have not been able to penetrate internally,
or operate any change in the eggs.   [A]

[A] Specific charaĉter of the Cock, *Phafianus-Gallus* :—" Has
" a compreffed earuncle on its top ; a double one on its cheek ; its
" ears naked, its tail compreffed and rifing." Linnæus reckons
up fourteen varieties : 1. The Common Cock, *Gallus domeſticus :*
2. The Copped Cock, *Gallus criſtatus* : 3. The Five-toed Cock,
Gallus pentadaĉtylus : 4. The Crifped Cock, *Gallus criſpus :* 5. The
Perfian Cock, or Rumkin, *Gallus ecaudatus :* 6. The Creeper or
Dwarf Cock, *Gallus pumilio :* 7. The Bantam Cock, *Gallus pu-
fillus :* 8. The Rough-footed Cock, *Gallus plumipes :* 9. The
Turkifh Cock, *Gallus Turcicus :* 10. The Padua Cock, *Gallus Pa-
tavinus :* 11. The Mozambic Cock, or Blackamoor, *Gallus Morio :*
12. The Black Cock, *Gallus niger :* 13. The Tuberous Cock,
Gallus tophaceus : 14. The Woolly Cock, *Gallus lanatus.*

The 12th and 13th varieties were difcovered by Gmelin and
Pallas : the former is a native of Perfia, and has a blaĉkifh fkin ;
the latter has a fwelling comb.

M

THE TURKY.

The T U R K E Y*

Le Dindon, Buff.
Meleagris-Gallopavo, Linn. Gmel. &c.
Gallina Indiana, Zuin.
Il Gallinacio, Cett.

IF the Cock be the moſt uſeful bird in our court-yards, the Turkey is the moſt diſtinguiſhed, by its bulk, by the ſhape of its head, and certain natural habits poſſeſſed by few ſpecies. Its head is very ſmall in proportion to its body, and is deſtitute of the uſual decoration; for it is almoſt entirely featherleſs, and, together with a part of the neck, is only covered with a bluiſh ſkin, befet with red fleſhy *papillæ* on the fore part of the neck, and whitiſh *papillæ* on the hind part of the head, with ſome ſmall ſtraggling black hairs, and a few feathers ſtill more rare on the arch of the neck, and which are thicker in the lower part, a circumſtance which has not been noticed by naturaliſts. From the baſe of the bill, a kind of red fleſhy caruncle falls

---

* As the Turkey was unknown before the diſcovery of America, it has no name in the ancient languages. The Spaniards called it *Pavon de las Indias*, i. e. *the Peacock of the Indies*, becauſe it ſpreads its tail like a Peacock. The Italians term it *Gallo d'India*; the Germans, *Indianiſch Han*; the Poles, *Indiyk*; and the Swedes, *Kalkon*.

looſely

loosely over a third part of the neck, which
at first sight appears single; but when examined
is found to be composed of a double membrane.
A fleshy protuberance, of a conical shape and
furrowed with deep transverse wrinkles, rises
from the bottom of the upper mandible.   This
protuberance is scarcely more than an inch long
in its natural state of contraction; that is, when
the Turkey, seeing no objects but those to which
it is accustomed, and feeling no inward agita-
tion, walks calmly seeking its food.   But, on any
unusual appearance, especially in the season of
love, this bird, which in its ordinary state is
humble and tame, swells with instant rage; its
head and neck become inflated, the conical protuberance expands, and descends two or three
inches lower than the bill, which it entirely covers.   All these fleshy parts assume at the same
time a deeper red; it bristles up the feathers on
its neck and back, spreads its tail like a fan, while
its wings drop and even trail on the ground.   In
this attitude, he sometimes struts around his female, making a dull sound, produced by the air
escaping from the breast through the bill, and
followed by a long gabbling noise.   Sometimes
he leaves his female to attack those who disturb
him.   In both these cases, his motions are
composed; but they become rapid the instant he
utters the dull sound which we have mentioned.   He vents a shrill scream, which every body
knows, which intermits from time to time, and
                                        which

which he may be made to repeat as often as one choofes, by whiftling, or by forming any fharp tones. He then begins again to wheel round, which action, according as it is directed to his female, or pointed at the object that has provoked his difpleafure, expreffes attachment or marks rage: and it is obferved, that his fits are the moft violent at the fight of red clothes; he is then inflamed and becomes furious; rufhes on the perfon, ftrikes with his bill, and exerts himfelf to the utmoft to remove an object whofe prefence he cannot endure.

It is a curious and very fingular fact, that the conical caruncle, which lengthens and is relaxed when the bird is agitated by the violence of paffion, is relaxed in the fame manner after death.

Some Turkies are white, others variegated with black and white, others with white and rufty yellow, others are of an uniform grey, which are the moft uncommon of all. But in the greater number the plumage verges on black, with a little white near the ends of the feathers: thofe which cover the back and the upper furface of the wings are fquare at the extremities; and among thofe of the rump, and even of the breaft, there are fome with rainbow colours, occafioned by the different rays being reflected according to the various degrees of incidence. As they grow older the tints become more gloffy, and the reflections more diverfified. Many people imagine that white Turkies are the hardi-

I 3   eft;

eft ; and, for this reafon, that breed is preferred
in fome provinces : there are numerous flocks in
Pertois in Champaign.

The naturalifts have reckoned twenty-eight
quills in each wing, and eighteen in the tail.
But what is much more ftriking, and what will
readily diftinguifh this fpecies from any other
yet known, is a lock of hard black hair, five or
fix inches long, which, in our temperate climates, rifes from the lower part of the neck in
the grown male Turkey on the fecond year,
and fometimes about the end of the firft; and
before it appears, the place where it emerges is
marked by a flefhy prominence. Linnæus fays,
that this hair does not fprout till the third year
in the Turkeys bred in Sweden. If the fact be
certain, it would follow that this production is
the flower in proportion to the rigour of the
climate; and indeed one of the principal effects
of cold is, to check every fort of growth. This
lock of hair is the foundation of the epithet of
bearded, *(pectore barbato,)* which has been applied to the Turkey; an expreffion in every
refpect improper, for it does not grow from the
breaft, but from the lower part of the neck;
and, befides, it is not fufficient that there are
hairs; they ought never to receive the name of
beard, unlefs they rife from the chin, as in
Edwards's bearded vulture.

We fhould form an inaccurate idea of the tail
of the Turkey-cock, if we imagined that all the

3   feathers

feathers of which it is compofed can equally be
fpread like a fan. In fact, he has two tails,
an upper and an under one; the firft confifts
of eighteen broad feathers inferted round the
rump, and which are erected when the bird
ftruts about; the fecond, or lower one, is
formed of others which are not fo broad, and
remains always in a horizontal pofition. The
male is alfo diftinguifhed by a fpur on each foot,
which is of various lengths, but always fhorter
and fofter than in common cocks.

In the female, not only the fpurs, and the lock of
hair hanging from the lower part of the neck, are
wanting, but alfo the conical caruncle is fhorter,
and admits of no extenfion; both this caruncle
barbil, and the glandulous flefh that fheaths
the head, are of a paler red; fhe is fmaller alfo,
and has lefs expreffion, lefs refolution, and lefs
vigour of action; her cry is only a plaintive
accent; fhe never ftirs but to feek food or to fly
before danger: finally, fhe cannot perform the
ftrutting evolutions, not becaufe fhe has not the
double tail of the male, but on account of the
want of the *elevator* mufcles which erect the
very large feathers that compofe the upper
fan.

In the male, as in the female, the noftrils are
fituated in the upper mandible; the ears are
placed behind the eyes, thickly covered, and, as
it were, darkened by a multitude of little divided
feathers, pointed in different directions.

It

It will readily be fuppofed, that the beft Cock
is the ftrongeft, the livelieft, and the moft
vigorous in all his movements. Five or fix
females may be entrufted to his care. If there
are feveral males, they will fight with each
other, but not with the furious obftinacy of ordinary cocks; thefe even attack Turkies which
are double their fize, and kill them in the combat. The fubjects of the contention are equally
compliant to the males of both fpecies, if, as
Sperling fays, the Turkey-cock, when deprived
of his females, pays his addreffes to the common hens; and the Turkey-hens, in the abfence of their males, offer their favours to the
ordinary cock, and eagerly folicit his potent
embrace *.

The battles which the Turkey-cocks fight
among themfelves are far lefs vigorous; the vanquifhed does not always fly from the field of
battle, and fometimes he is even preferred by
the females. It has been obferved, that though
a white Turkey was beat by a black one, all the
chickens were white.

The Turkies perform copulation nearly in
the fame way as ordinary cocks, only it lafts
longer. Hence, perhaps, the reafon that the
male is not equal to fo many females, and is
fooner worn out. I have already mentioned,
on the authority of Sperling, that he fometimes

* Zoologia Phyfica, p. 367.

mixes

mixes with common hens; the fame author afferts, that when his females are taken from him, he not only couples with the pea-hen (which may happen), but alfo with the ducks (which feems to me to be lefs probable).

The Turkey-hen is not fo prolific as the common hen. She muft, from time to time, be fed with hemp-feed, oats, and buck-wheat, to make her lay: and after all, fhe feldom has more than one hatch of fifteen eggs a-year. When fhe has two, which is very uncommon, the firft is about the end of winter, and the fecond in the month of Auguft. The eggs are white, with fome fmall fpots of reddifh-yellow; and their ftructure is nearly the fame as in thofe of the common hen. The Turkey-hen will alfo hatch the eggs of all forts of birds. We may know when fhe wants to fit, for fhe remains in the neft; and in order to fix her attachment, the place muft be dry, with a good afpect, according to the feafon, and not too much expofed; for inftinct leads her to conceal herfelf with the greateft care when fhe covers.

Thofe of a year old are generally the beft fitters, and they are fo intent, that they would die upon their eggs from mere inanition, if we were not at pains to remove them once a-day, and give them food and drink. This paffion for hatching is fo powerful and fo durable, that they fometimes have two nefts in fucceffion, without the leaft interruption; but, in

fuch

such cases, they must be supported by richer food. The cock has a very opposite instinct; for if he sees the female covering, he breaks the eggs, which he regards, probably, as an obstacle to his pleasures * ; and for this reason it is, perhaps, that the female is so industrious in concealing her nest.

After the full time, when the young Turkies are about to burst into day, they pierce with their bill the shell of the egg in which they are inclosed ; but it is sometimes so hard, that they would perish if not assisted by crushing it ; and this must be effected with great caution, following as closely as possible the natural process. If roughly handled in their tender moments, if suffered to endure hunger, or if exposed to inclement weather, though they may survive for the time, they will pine away and soon perish. Cold, rain, and even dew, occasions lingering sickness; the rays of a bright sun strike them with instant death; and sometimes they are crushed even under the feet of their mother: such are the dangers which threaten the life of this delicate bird. This cause, joined to the inferior fecundity of the Turkey-hens in Europe, is the reason why this species is much less numerous than that of the common poultry.

After their extrication from the shell, the young Turkies ought to be kept in a warm and

* Sperling. *Ibid.*

dry

dry place, where there is spread a broad layer of dung well trodden; and when we would bring them out into the open air, we should do it by degrees, and chuse the finest days.

The young Turkies instinctively prefer picking out of the hand, to feeding in any other way. We judge by their chirping when they want to eat, which is frequent. They ought to be presented with food four or five times a-day; their first nourishment should be wine and water, which must be blown into their bill, and afterwards a few crumbs of bread may be mixed with it. On the fourth day, eggs spoiled in hatching may be given, beat up with bits of bread; and these addled eggs, whether they be hens or Turkies, are found to afford a salutary nourishment*. Towards the tenth or twelfth day, the eggs are omitted, and in their stead are used chopped nettles mixed with millet, or with the flour of Turkey beans, of barley, of wheat, or of buck-wheat; or at least, if we would save the grain without hurting the brood, with curdled milk, burdock, a little stinking camomile, nettle-feed, and bran. Afterwards, it will be sufficient to give them all sorts of decayed fruits cut into bits †, especially the berries of brambles and of white mulberries, &c. When we perceive them having a languishing appear-

* Journal Economique, *Aôut* 1757, p. 69—73.
† *Id. ibidem.*

ance,

ance, we muſt dip their bills into wine, to make
them drink a little, and at the ſame time oblige
them to ſwallow a grain of pepper. Sometimes
they appear benumbed and motionleſs, when
they have been overtaken by a cold rain; and
they would infallibly die, if we were not careful
to wrap them in warm rags, and blow repeatedly
into them warm air through their bill. They
muſt be viſited from time to time to pierce the
ſmall bladders that collect under the tongue and
about the rump, and to give them ruſt-water;
it is even recommended to bathe their head with
this water, to prevent certain diſeaſes to which
they are ſubject; but in that caſe, it muſt be
wiped and dried very carefully; for it is well
known that humidity of every kind is hurtful to
Turkies in their tender age.

The mother leads them with the ſame ſolicitude that the hen leads her chickens; ſhe warms
them under her wings with the ſame affection,
and protects them with the ſame courage. It
would ſeem that tenderneſs for her offspring
gives quickneſs to her ſight; ſhe deſcries a bird
of prey at a prodigious diſtance, when it is yet
inviſible to every other eye. As ſoon as ſhe
perceives her dreaded enemy, ſhe vents her fears
by a ſcream that ſpreads terror through the
whole brood; each little Turkey ſeeks refuge
under a buſh, or ſquats in the herbage, and the
mother keeps them in that ſituation by her cries

ſo

fo long as danger is impending; but when her apprehenfions are removed, fhe informs them by a different note, and calls them from their concealments to affemble round her.

When the Turkies are newly hatched their head is fhaded with a kind of down, but they have ftill no glandulous flefh or barbils. It requires fix weeks or two months till thefe parts develope, or, as it is vulgarly faid, before the Turkies *put forth the red* *. This is as critical a period with them as that of dentition is with children; and then efpecially wine ought to be mixed with their ordinary food to ftrengthen them. A fhort while before this they have begun to perch.

It is feldom that Turkies are fubjected to caftration as ordinary cocks are; they fatten very well without fuffering that operation, and their flefh is no lefs delicate: another proof that their temperament is not fo hot as that of common poultry.

When they have grown hardy, they leave their mother, or rather they are abandoned by her. The more tender and delicate they were in their infancy, they become in time the more robuft and the more capable of fupporting the inclemency of the weather. They are fond of perching in open air, and thus pafs whole nights in the rigours of winter; fometimes refting on

* *Pouffer le rouge.*

one

one foot and drawing up the other to keep it warm, as it were, under the ventral feathers; at other times, on the contrary, crouching on the branch, and keeping their bodies in equilibrium. They lay their head under the wing when they go to ſleep, and, during their repoſe, the motion of reſpiration is very perceptible.

The beſt way of training Turkies after they are grown ſtout is, to allow them to ramble in the fields where nettles, and other plants which they are fond of, are plentiful, or to admit them into the orchards when the fruit begins to drop, &c. But we muſt be attentive to reſtrain them from thoſe paſtures that bear plants hurtful to them, ſuch as the great fox-glove with red flowers; this plant is a real poiſon to Turkies; thoſe that eat it are thrown into a kind of intoxication, vertigoes, and convulſions, and when the quantity is large they languiſh and die. We cannot therefore be too careful in extirpating this noxious plant from thoſe places where Turkies are raiſed *.

We ſhould alſo be careful, eſpecially in their early infancy, not to ſuffer them to go abroad in the morning till the ſun has dried the dew, and to ſhut them up before the fall of the evening damps; they muſt likewiſe be confined

* Hiſtoire de l'Academie Royale des Sciences des Paris, année 1748, p. 84.

in

in the fhade during the violent heats of the fummer's day. Each evening, when they return to
rooft, they muft be fed on pafte of grain, or on
fome other food, except in harveft, when they
have gathered enough in the fields. As they
are extremely timid, they are eafily led; the
very fhadow of a fwitch is fufficient to drive
large flocks, and they will often run from an
animal that is much fmaller and much weaker
than themfelves. There are occafions, however, when they difcover courage, efpecially in
their defence againft the affaults of pole-cats,
and other foes of the poultry. Sometimes even
they furround a hare in his feat, and ftrive to
kill him by ftriking with their bill *.

They have different tones, and different inflexions of voice, according to their age, their
fex, and the various paffions by which they are
influenced; their pace is flow, their flight tardy;
they drink, eat, and fwallow fmall pebbles
nearly as the cocks do, and have alfo a double
ftomach, that is, a craw and a gizzard ; but, as
they are much larger, the mufcles of the gizzard
are alfo much ftronger.

The length of the inteftines is nearly quadruple that of the bird, reckoning from the tip
of the bill to the end of the rump; they have
two *cæca*, both turned forwards, and which,
taken together, conftitute more than a fourth of

* Salerne.

the

the whole alimentary canal; thefe take their
rife near the extremity, and the excrements
contained in their cavity differ but little from
thofe included within the *colon* and *rectum*; thefe
excrements do not remain at all in the common
cloaca, as the urine, and that white fediment
which is always found where the urine paffes,
and they have confiftence enough to receive
fhape in their extrufion from the *anus*.

The parts of generation are nearly the fame
in Turkies as in common cocks; but they feem
to be much lefs vigorous in their performance.
The males are not fo ardent for the females;
their embraces are lefs frequent and lefs expedi-
tious; and the females, at leaft in our climate,
lay much later, and have much fewer eggs.

As the eyes of birds have in fome refpects a
different organization from thofe of man and of
quadrupeds, it may be proper to mention the
chief diftinctions. Befides the upper and under
eye-lids, the Turkies, as well as moft other
birds, have ftill a third, called the internal eye-
lid, *membrana nictitans*, which draws itfelf back
into the fhape of a crefcent in the large angle of
the eye, and whofe quick and frequent twink-
lings are effected by a curious mufcular con-
trivance. The upper eye-lid is almoft entirely
immoveable, but the lower can fhut the eye by
rifing to the upper, which fcarcely ever happens, except when the animal is afleep. Thefe
two eye-lids have each a lachrymal point, but

no

no cartilaginous margins; the *cornea* is tranfparent, encircled by an offeous ring, confifting
of about fifteen fcales over-lapping each other
like the tiles of a roof. The cryftalline lens is
harder than in man, but fofter than in the
quadrupeds or fifhes *, and its pofterior furface
is the moft convex †. Laftly, the optic nerve
fends off, between the retina and the choröid
coat, a black membrane of a rhomboidal figure,
confifting of parallel fibres, which ftretch through
the vitreous humour, and are attached fometimes directly to the interior angle of the cryftalline capfule, and fometimes are connected by
the intervention of a filament branching from it.
It is to this fubtile and tranfparent membrane
that the academicians have given the name of
purfe, though it has fcarcely any refemblance to
that in the Turkey, and ftill lefs in the cock,
the goofe, the duck, the pigeon, &c.; and its
ufe, according to Petit, is to abforb the rays of
light that come from the lateral objects, and
which, intermingling with the others, would
render vifion obfcure ‡. However this may be,
certain it is, that the organ of fight is more
complex in birds than in quadrupeds; and as
we have before fhewn that this fenfe is poffeffed
by the feathered race in a higher degree than
what obtains in other animals, we muft

* Memoires de l'Academie Royale des Sciences, *année* 1726.
p. 83.

† *Id. année* 1730. p. 10.   ‡ *Id. année* 1735, p. 123.

afcribe the fuperiority to its difference of ftructure, and to its more perfect organization; but
to ftate the precife effect would require a more
profound ftudy of comparative anatomy and of
the animal œconomy.

If we compare the relations of travellers, we
cannot hefitate to conclude, that Turkies are natives of America and of the adjacent iflands;
and that before the difcovery of the New Continent, there exifted none in the Old.

Father du Tertre obferves, that the Antilles are
their congenial abode ; and that, if a little care
be beftowed, they will there hatch three or four
times in the year*. But it is a general principle, that all animals multiply fafteft and grow
largeft and ftouteft in their original refidence:
and this is exactly what takes place with regard
to the Turkies in America. Immenfe numbers,
we are told by the Jefuit miffionaries, inhabit near
the river Illinois † ; flocks of an hundred, fometimes even of two hundred, are feen at once. They
are much larger than thofe in Europe, and weigh
even thirty-fix pounds: Joffelin affirms, that fome
are fixty pounds ‡. They are no lefs plentiful
in Canada (where, according to Fathers Theodat
and Recollet, the favages call them *Ondettouta-
ques*), in Mexico, in New England, in the vaft
country watered by the Miffiffippi, and in the

* Hiftoire Generale des Antilles, *tome* ii. p. 266.
† Lettres Edifiantes, xxiii. p. 237.
‡ New England Rarities, p. 8.

Brazils,

Brazils, where they pafs by the name of *Arig-
nanouſſou* *. Dr. Hans Sloane faw fome in Jamaica; and he remarks, that in almoſt all thefe
countries they are in the wild ſtate, and fwarm
at fome diſtance from the plantations, and but
gradually retire from the encroachments of the
European fettlers.

But if moſt travellers and obfervers agree that
Turkies are natives of America, efpecially of
the northern part of that continent, they are no
lefs unanimous in opinion that there are extremely few or none of thefe birds in the whole of
Afia.

Gemelli Careri informs us, that not only there
were none originally in the Philippine Iſlands,
but that thofe introduced by the Spaniards from
Mexico did not thrive.

Father du Halde aſſures us, that none are to be
found in the empire of China, except what have
been carried thither †. It is true, indeed, that
this Jefuit fuppofes in the fame place, that thefe
birds are common in the Eaſt Indies; but it
would feem that this is only a fuppofition founded on report; whereas he was an eye-witnefs of
the fact that he relates with refpect to China.

Father de Bourzes, another Jefuit, fays, that
there are none in the kingdom of Madura, which
lies in the peninfula on this fide of the Ganges;
and he therefore concludes, with probability, that

---

* Voyage au Brefil, recuelli par de Lery, p. 171.
† Hiſtoire Generale des Voyages, *tome* vi p. 487.

it

it is the West Indies that have given name to this
bird *.

Dampier saw none at Mindanao †; Chardin
and Tavernier, who travelled over Asia, affirm
positively, that there are no Turkies in the whole
of that vast country ‡. According to Tavernier,
the Armenians introduced them into Persia,
where however they have not succeeded well;
the Dutch carried them to Batavia, where they
have thrived exceedingly.

Finally, Bosman and some other travellers tell
us, that if Turkies be ever seen in the country
of Congo, on the Gold Coast, at Senegal, or
in other parts of Africa, it is only at the
factories and with strangers, the natives making little use of them. According to the same
travellers, their Turkies are evidently descended
from those carried thither by the Portuguese and
other Europeans, along with other poultry ‖.

I will not dissemble, that Aldrovandus, Gesner, Belon, and Ray, have affirmed that Turkies
were natives of Africa or of the East Indies; and
though their opinion on this subject is at present
little regarded, I conceive that it is a duty which I
owe to these great names not to reject it without some discussion.

* *Indian Cock.* See his letter of the 21st September 1713,
among the *Lettres Edifiantes*.

† New Voyage, vol. i.

‡ Voyages de Chardin, *tome* ii. p. 29. Voyages de Tavernier,
tome ii. p. 22.

‖ Bosman.

7   Aldrovan-

Aldrovandus has attempted to prove at great length, that Turkies are the *Meleagrides* of the ancients, or the African or Numidian Hens, whofe plumage was covered with round fpots, like drops *(Gallinæ Numidicæ guttatæ)*; but it is evident, and every body is now agreed on the fubject, that thefe are really our *pintados*, which indeed come from Africa, but which are quite different from Turkies. It would therefore be needlefs to dwell more particularly on the opinion of Aldrovandus, which in fact carries its refutation along with it; and yet Linnæus feems inclined to perpetuate or renew the error, by applying to the Turkey the name of *Meleagris*.

Ray, who maintains that Turkies have derived their origin from Africa or the Eaft Indies, feems to have fuffered himfelf to be deceived by names. That of the bird of *Numidia*, which he adopts, implies an African defcent; that of *Turkey* and the *Bird of Calecut*, denotes an Afiatic extraction. But no proof can be drawn from the name beftowed by ill-informed people, or even the fcientific term impofed by philofophers, who are not always exempted from prejudices. Befides, Ray himfelf admits with Dr. Sloane, that thefe birds delight in the warm countries of America, and there multiply prodigioufly *.

With regard to Gefner, he admits indeed that moft of the ancients, and among others Ariftotle

---

* Synopfis Avium, Append. p. 182.

and

and Pliny, were totally unacquainted with Turkies; but he fuppofes that in the following quotation Ælian had them in view : " India," fays
this ancient, " produces a fort of very large cocks,
"₁ whofe comb is not red like that of ours, but
" fo rich and variegated as to refemble a crown
" of flowers; the feathers of the tail are not
" arched nor bent into circles, but flat, and when
" they are not erected, they trail like thofe of
" the peacock; their plumage is of an emerald
" colour."

But it does not appear that this paffage relates
to Turkey Cocks; for, 1. The fize does not
prove the point, it being well known that in
Afia, and efpecially in Perfia and at Pegu, the
common cocks are exceedingly large.

2. This comb, compofed of various colours, is
alone fufficient to overturn the opinion, fince
Turkies have never any comb; and what is here
defcribed is not a tuft of feathers, but a real
comb, fimilar to that of the ordinary cock, though
of a different colour.

3. The manner it holds its tail, refembling the
peacock, is equally inconclufive; for Ælian pofitively fays, that the bird which he is defcribing
carries its tail like the peacock, *when it does not
erect it ;* and if there had been an erection, accompanied with a wheeling motion, Ælian would
not have omitted a character fo fingular, and
which forms fo ftriking a refemblance to the peacock,

cock, with which he was at the fame inftant drawing a comparifon.

4. Laftly, The emerald colour of the plumage is not fufficient to decide whether the defcription refers to the Turkey, though fome of its feathers have that tinge, and in certain fituations reflect that fort of light, fince many other birds have the fame properties.

Nor does Belon feem to reft his opinion on better foundation, when he attempts to difcover Turkies in the writings of the ancients. Columella had faid in his treatife *De re ruftica : Africana eft meleagridi fimilis, nifi quod rutilam galeam et criftam capite gerit, quæ utraque in meleagrida funt cærulea**. "The African hen is "like the meleagris, only its tuft and comb are "red, but in the other both are cœrulean." Belon takes this *African hen* for the pintado, and the meleagris for the Turkey; but it is evident from the paffage itfelf, that Columella fpeaks here only of two varieties of the fame fpecies; fince the two birds mentioned are perfectly alike, except in colour, which is liable to vary in the fame fpecies, efpecially in that of the pintado, of which in the males the membranous appendices that hang on both fides of the cheeks are of a blue colour, while in the female the fame parts are red. Befides, is it likely that Columella, wifhing to diftinguifh two fpecies fo remote from each other as the pintado and the Turkey,

* Lib. viii. 2.

would

would be contented in felecting a flight difference of colour, inftead of marking obvious and
ftriking characters?

But if the attempts of Belon to beftow on
Turkies, from the authority of Columella, an
African origin, are without foundation, his
fuccefs is not greater, when he feeks, from the
following paffage of Ptolemy, to give them an
Afiatic origin :—" Trigliphon Regia, where
" the common cocks are faid to be bearded."
This Trigliphon is fituated indeed beyond
the Ganges; but there is no reafon to believe that thefe bearded cocks are Turkies;
for, 1. The very exiftence of thefe cocks is
uncertain, refting merely on hearfay. 2. This
defcription cannot refer to Turkies, fince, as I
have before obferved, the word *beard*, applied
to a bird, can mean only a tuft of feathers, or
hairs, placed under the bill, not the lock of ftiff
hair which the Turkies have on the under part
of the neck. 3. Ptolemy was an aftronomer
and geographer, and not a naturalift; and it is
evident that he wifhed to render his charts more
interefting, by introducing, and not always with
judgment, accounts of the peculiarities of each
country. In the very fame page he fpeaks of
three iflands of fatyrs, whofe inhabitants had
tails; and he tells us, that the Manioles are ten
iflands fituated nearly in the fame climate,
where loadftone abounds fo much, that iron
cannot be employed in the conftruction of fhips,

becaufe

becaufe of the danger of their being attracted and held by the magnetic force. But thefe human tails, though afferted by feveral travellers, and by the Jefuit miffionaries, according to Gemelli Careri, are at leaft very doubtful; and the mountains of loadftone, or rather their effects on the iron of veffels, are no lefs fo: fo that little confidence can be put in facts mingled with fuch uncertain relations. 4. Laftly, Ptolemy, in the place above quoted, fpeaks exprefsly of ordinary cocks, which cannot be confounded with Turkey-cocks, neither in their external form, their plumage, their cry, their natural habits, the colour of their eggs, nor the time of incubation, &c. It is true that Scaliger, while he admits that the meleagris of Athenæus, or rather of Clytus, who is quoted by Athenæus, was an Ætolian bird that loved wet fituations, which was averfe to hatching, and whofe flefh had a marfhy tafte, none of which characters belong to the Turkey; which is not an inhabitant of Ætolia, which avoids watery fpots, which has the greateft affection to its young, and whofe flefh has a delicate flavour; yet ftill maintains that the meleagris is the Turkey. But the anatomifts of the Academy of Sciences, who were at firft of the fame opinion, have, after examining the fubject with more attention, afcertained and proved that the pintado was the real *meleagris* of the ancients. In fhort, we

muft

muft confider it as an eftablifhed point, that Athenæus, or Clytus, Ælian, Columella, and Ptolemy, have no more fpoken of Turkies than Ariftotle or Pliny; and that thefe birds were totally unknown to the ancients.

Nor can we find the leaft mention of the Turkey in any modern work, written prior to the difcovery of America. A popular tradition refers the period of its firft introduction into France to the fixteenth century, in the reign of Francis I.; for this was the time when Admiral Chabot lived. The authors of the Britifh Zoology ftate it as a well-known fact, that they were introduced into England in the time of Henry VIII. the contemporary of Francis I.; which agrees exactly with our opinion. For America having been difcovered by Chrifto pher Columbus towards the end of the fifteenth century, and thefe fovereigns having afcended the throne about the beginning of the fixteenth century, it is natural to fuppofe, that the Turkies brought from the New World would under their reigns be regarded as novelties in France and England. This is confirmed too by the exprefs teftimony of J. Sperling, who wrote before 1660; he affirms, that they had been introduced from the New Indies into Europe more than a century prior to his time *.

Every thing, therefore, concurs to prove that the Turkies are natives of America. As they are

* Zoologia Phyfica, p. 366.

heavy

heavy birds, and cannot rife on the wing, or fwim, it would be impoffible for them to crofs the ocean which feparates the two continents. They are in the fame fituation with the quadrupeds, which, without the affiftance of man, would not have been difperfed through the Old and New Worlds. This reflection gives additional weight to the teftimony of travellers, who affure us, that they have never feen Wild Turkies either in Africa or Afia, and that none are found there but fuch as are domeftic, and brought from other parts *.

This

* The Honourable Daines Barrington has publifhed an Effay, in which he attempts to prove, that the Turkey was known before the difcovery of America. He examines the Comte de Buffon's arguments, and endeavours to invalidate or refute them ; but his objections are entirely inconclufive. If the Turkey had been introduced into Spain by Columbus, it would have been called, fays Mr. Barrington, *the Mexican bird*, and not *pago*, or *pavo*. Cardinal Perron, who died in 1620, relates, that the Indian Cocks had prodigioufly multiplied, and were driven like flocks of fheep from Languedoc into Spain : therefore, fays Mr. B. the Turkey muft have been introduced firft into France. Thefe conjectures are fo vague as to merit no particular difcuffion ; and when Mr. B. afferts, that Sperling means one hundred and one years, by the expreffion " Ante centum, & *quod excurrit* annos," he feems not to have attended to grammar. That phrafe is claffical, and means indefinitely fome time more than a century : nor will the word *excurrit* admit of any other interpretation.

Mr. Barrington proceeds : " The Spanifh term is not *pavon* " *de las Indias*, as M. Buffon ftates, but fimply *pavo*, and for- " merly *pago*. If, moreover, the name were *pavon de las Indias*, " it would not fignify the Weft Indies, as in all the European lan- " guages the addition of *Weftern* is neceffary." But this affertion is too hafty : did not the King of Spain, after the difcovery of

America,

This determination of the natal region of the
Turkey leads to the decifion of another queftion,
which, at firft fight, feems to have no connection with it.   J. Sperling affirms, in his *Zoo-
logica Phyfica*, p. 369, that the Turkey. in a
monfter (he means an hybrid) produced by the
union of the two fpecies, that of the peacock
and of the ordinary cock; but as it is afcertained that the Turkey is of American extraction, it could not be bred by the intercourfe of two Afiatic fpecies; and what completely decides the point is, that no Wild Turkies
are found through the whole extent of Afia,
though they abound in the forefts of America.
But it will be faid, what means the term *gallo-
pavus*, which has fo long been applied to the
Turkey? Nothing is fimpler: the Turkey was
a foreign bird which had no name in any of
the European languages; and as it bore feveral
ftriking refemblances to the common cock and
the peacock, a compound word was formed
expreffive of thefe analogies.   Sperling and
others would have us believe that it is really the
crofs-breed of thefe two fpecies; yet the inter-

America, affume the title of *Indiæ Rex,* and not *Indiæ Occidenta-
lis,* or *Indiarum?*

I muft add, that the opinion of the Comte de Buffon concerning the native climate of the Turkey, is admitted by the ingenious and refpectable naturalift Mr. Pennant, who has adduced
feveral new arguments in fupport of it.   Linnæus, Gmelin, and
Latham, entertain the fame idea.

mixture

mixture confifts entirely in the names.—So
dangerous it is to beftow upon animals compounded epithets, which are always ambi-
guous.

Edwards mentions another hybrid produced
between the Turkey and the pheafant *.
The individual which he defcribes was fhot
in the woods near Hanford in Dorfetfhire,
where it was feen in the month of October
1759, with two or three other birds of the fame
kind. It was of a middle-fize between the
pheafant and the Turkey, its wings extending
thirty-two inches; a fmall tuft of pretty long
black feathers rofe on the bafe of the upper
mandible; the head was not bare like that of
the Turkey, but covered with little fhort feathers; the eyes were furrounded with a circle
of red fkin, but not fo broad as in the pheafant.
It is not faid whether this bird could fpread the
large feathers of the tail into the wheel-fhape;
it only appears from the figure, that it carried
the tail in the fame way as the Turkey generally
does. It muft alfo be obferved, that this tail
is compofed of fixteen quills, as in the grous;
while that of the Turkey and of the pheafant
confifts of eighteen; alfo each feather on the
body fhot double.from the fame root, the one
branch ftiff and broad, the other fmall and covered with down, a character which belongs

* Gleanings.

neither

neither to the pheafant nor the Turkey. If this
bird was really a hybrid, it ought to have had,
like other hybrids, 1ft, The characters common
to the two primitive fpecies ; 2dly, The qualities
intermediate between the extremes ; a circumftance that in this cafe does not take place, fince
this individual had a character not to be found
in either (the double feathers), and wanted
others that occur in both (the eighteen quills of
the tail). Indeed, if it be infifted that it was
hybridous, we fhould more reafonably infer,
that it was produced by the union of the Turkey
with the grous; which, as I have remarked,
has no more than fixteen feathers in the tail,
but has the double feathers.

The Wild Turkies differ not from the domeftic fort, except that they are much larger
and blacker ; they have the fame difpofitions, the
fame natural habits, and the fame ftupidity.
They perch in the woods on the dry branches,
and when one falls by a fhot, the reft are not
intimidated by the report, but all continue fecure
in the fame pofition. According to Fernandes,
their flefh, though pleafant to eat, is harder and
not fo delicate as that of the Tame Turkies; but
they are twice as large. The Mexican name
of the male is *hucxolotl*, and that of the female
cihuatotolin. Albin tells us, that many Englifh
gentlemen amufe themfelves in breeding Wild
Turkies, and that thefe birds thrive very well
in fmall woods, parks, or other inclofures.

The

The Crefted Turkey is only a variety of the common kind, fimilar to what occurs among the ordinary cocks. It is fometimes black, fometimes white. That defcribed by Albin was of the ufual fize; its feet flefh-coloured, the upper part of the body deep brown; the breaft, belly, thighs, and tail, white; and alfo the feathers that form the tuft. In other refpects it refembled exactly the ordinary kind; it had the fpongy and glandulous flefh which covers the head and arch of the neck, and the lock of hard hair rifing (apparently) from the breaft, and the fhort fpurs on each foot; it alfo bore the fame fingular antipathy to fcarlet, &c. [A]

[A] Specific character of the Turkey, *Meleagris Gallopavo :* " The caruncle of the head is extended to the forehead and the " throat; the breaft of the male is bearded." The Wild Turkies are of a dingy uniform colour; and feldom weigh more than thirty pounds. They are now very rare in the old fettlements of North America; yet fome occur in Virginia within 150 miles of the roaft. Beyond the ridge of Apalachian mountains they are frequent; and flocks of feveral hundreds are feen near the Miffiffippi and Ohio. They rooft in the great fwamps, but fpend the whole day among the dry woods, fearching for red acorns and various forts of berries. They grow very fat in the fpring. When furprifed, they run with prodigious fpeed; but if hotly purfued, they take wing and perch on the fummit of the next tall tree. The Indians make fans of the Turkies tails; and alfo weave the inner webs of their feathers with hemp, or the rind of the mulberry-tree, into an elegant fort of clothing.

M

The GUINEA PINTADO *.

La Peintade, Buff.
Numida Meleagris, Linn. Gmel. &c.
Gallus & Gallina Guineenfis, Ray and Will.
The Guinea Hen, Ray.

WE muſt not, like Ray, confound this with the Pintado mentioned by Dampier, which is a ſea-bird, equal to the duck in ſize, having very long wings, and ſkimming along the ſurface of the water: theſe charaċters are all widely different from thoſe of the real Pintado, which is a land-bird, with ſhort wings, and whoſe flight is laborious and ſlow.

It was known, and accurately deſcribed, by the ancients. Ariſtotle mentions it only once in his Hiſtory of Animals; he calls it *Meleagris*, and ſays that its eggs are marked with ſmall ſpots †.

Varro takes notice of it by the name of African Hen; and he tells us, that it was a large bird,

---

* In Greek and Latin, *Meleagris*: in modern Italian, *Gallina di Numidia*: in German, *Perl-buhn*, or *Pearl-hen*. In Congo it has the name *Quetelé*.

† Lib. vi. 11.

with

THE GUINEA PINTADO .

with variegated plumage, and a round back,
that was very uncommon at Rome *.

Pliny gives the same account, and seems
merely to copy Varro; unless we ascribe the
sameness of their descriptions to the identity of
the object †. He repeats also what Aristotle
had said with regard to their eggs ‡; and he
adds, that the Pintado of Numidia was most
esteemed §, and hence he bestows the name of
Numidian Hen on the whole species.

Columella admitted two sorts, which were
perfectly alike, except that the one had blue
barbils and the other red. This difference
seemed so important to the ancients, that they
formed two species, denoted by distinct names.
They called the one, which had red barbils,
Meleagris; the other, which had blue barbils, the *African Hen* ‖; not adverting that the
former is the female, and the latter the male
of the same identical species, as the academicians have found ¶.

---

* *Grandes, variæ, gibberæ quas Meleagrides appellant Græci.*
Lib. iii. 9.

† *Africæ Gallinarum genus, gibberum, variis sparsum plumis.*
Lib. x. 26.

‡ Lib. x. 52.

§ Lib. x. 48. " *quam plerique Numidicam dicunt.*"

‖ *Africana Gallina est Meleagridi similis nisi quod rutilam paleam
& cristam capite gerit, quæ utraque sunt in Meleagrida cærulea.*
COLUMELLA *de Re Rustica*, lib. xiii. 2.

¶ Memoires pour servir a l'Histoire Naturelle des Animaux,
dressé par M. Perrault. *Deuxieme Partie*, p. 82.

How-

However, it appears that the Pintado which was anciently reared with fo much care at Rome, was afterwards entirely loft in Europe. We can difcover no trace of it in the writings of the middle ages; and we find it only begun to be fpoken of, after the Europeans had vifited the weftern coafts of Africa, in their voyages to India by the Cape of Good Hope *. But not only have they diffufed thefe birds through Europe, but tranfported them into America; and the Pintados have fuffered various alterations in their external qualities from the influence of different climates. Nor muft we be furprifed that the moderns, both the naturalifts and travellers, have multiplied the divifions of the breeds ftill more than the ancients.

Frifch diftinguifhes, like Columella, the Pintado with red barbils from that with blue barbils; but he ftates feveral other differences. According to him, the latter, which is found only in Italy, is unpalatable food, fmall, fond of wet places, and carelefs about its young; the two laft features alfo mark the *Mcleagris* mentioned by Clytus of Miletus. " They delight," fays he, " in marfhes, and difcover little attach-

* " As Guinea is a country from which merchants have im-
" ported many articles formerly unknown to the French, fo its
" hens would alfo have remained unknown, had they not been
" brought over fea. But they are now fo frequently kept by the
" great lords in our provinces, as to be reckoned common."

BELON.

" ment

" ment to their progeny, which the priests are
" obliged to watch over with care;"—" but," he
subjoins, " their size is equal to that of a hen of
" the finest breed *." It appears too, from
Pliny, that this naturalist considered the Meleagris as an aquatic bird †. That with red barbils is, on the contrary, according to Frisch,
larger than a pheasant, prefers a dry situation,
and is assiduous in its attention to its young,
&c.

Dampier informs us, that in the island of
May, one of the Cape de Verd islands, there are
Pintados, of which the flesh is of an uncommon whiteness; and others, of which it is
black; but that in all of them it is tender and
delicate. Father Labat gives the same account.
This difference, if the fact be true, would appear to be the more considerable, as it cannot
be imputed to the change of climate; since the
Pintados on this island, which is near the African shore, may be considered as in their native
residence; at least unless we assert that the
same causes which tinge with black the skin and
periosteum of most of the birds in the islands of
St. Jago, darken also the flesh of the Pintados
in the neighbouring island of May.

* See Athenæus, lib. xiv. 26.
† " Mnesias calls a place in Africa, Sicyone; and a river,
" Crathis; which rises out of a lake where the birds termed
" *Meleagrides & Penelopæ* haunt." Lib. xxxvii. 2.

Father

Father Charlevoix pretends that there is at St. Domingo a fpecies fmaller than the ordinary fort *. But thefe are probably the chefnut Pintados, bred from fuch as were introduced by the Caftilians foon after the conqueft of the ifland. Thefe having become wild, and as it were naturalized in the country, have experienced the baneful influence of that climate; which, as I have elfewhere fhewn, has a tendency to enfeeble, to contract, and to degrade the animal tribes. It is worth obferving, that this breed, originally from Guinea, and tranfported to America, where it had once been reduced to the domeftic ftate, but fuffered to grow wild, could not afterwards be reclaimed to its former condition; and that the planters in St. Domingo have been obliged to import tame ones from Africa, to propagate in their farmyards †. Is it from living in a more defert and wilder country, inhabited by favages, that the chefnut Pintados have become favage themfelves? or is it becaufe they have been frighted away by European hunters, efpecially the French, who, according to Father Margat the Jefuit, have deftroyed vaft numbers of them ‡?

Marcgrave faw fome with crefts, that came from Sierra Leona, and which had about their

* Hiftory of the Spanifh ifland of St. Domingo.
† Lettres Edifiantes, xx.   ‡ *Ibidem.*

neck

neck a kind of membranous collar, of a bluifh cinereous colour *; and this is one of thofe varieties which I call primitive, and which deferve the more attention, as they are anterior to every change of climate.

The Jefuit Margat, who admits no fpecial difference between the African Hen and the Meleagris of the ancients, fays, that they have two kinds in regard to colour at St. Domingo; in fome, black and white fpots are difpofed in the form of rhomboids; in others, the plumage is of a deeper afh-grey. He adds, that they all have white below the belly, and on the underfide, and at the tips of the wings.

Laftly, Briffon confiders the whitenefs of the plumage of the breaft obferved on the Pintados at Jamaica, as conftituting a diftinct variety; and he characterifes it by this epithet, *(pectore albo;)* which, as we have juft feen, belongs as much to the Pintados of St. Domingo as to thofe of Jamaica.

But befides the differences which have been regarded by naturalifts as a fufficient foundation for admitting feveral races of Pintados, I can perceive many others, in comparing the defcriptions and figures publifhed by different authors, which fhew little permanency, either in the

---

* " The head was covered with a roundifh creft, much divided, and confifting of elegant black feathers."

Hift. Naturalis Brafilienfis.

L 3   internal

internal mould of the bird, or in the impreffion of the exterior form; but, on the contrary, a great difpofition to be affected by foreign influences.

In the Pintado of Frifch and fome others *, the cafque and the feet are whitifh, the forehead, the circle of the eyes, the fides of the head and neck, in its upper part, are white, fpotted with afh-grey.   That of Frifch has befides, under the throat, a red fpot in the fhape of a crefcent, and lower down a very broad black collar, the filky filaments on the *occiput* few, and not a fingle white quill in the wings; which form fo many diverfities, in which the Pintados of thefe authors differ from ours.

In Marcgrave's fpecimen, the bill was yellow; in that of Briffon, it was red at the bafe, horn-coloured near the tip.   The academicians found on fome a fmall tuft at the origin of the beak, confifting of twelve or fifteen ftiff threads, about four inches long, which did not occur in thofe of Sierra Leonà, mentioned above.

Dr. Caius fays, that in the female the head is entirely black, and that this is the only diftinction between it and the male †.

---

* " The cock and hen," fays Belon, " have the fame marbling on the feathers, and whitenefs about the eyes, and rednefs below." " At the fides of the head white." MARCGRAVE.— " The head is clothed," fays the Jefuit Margat, " by a fpongy, rough, wrinkled fkin, whofe colour is whitifh blue."

† *Apud Gefnerum.*

Aldrovandus

Aldrovandus afferts, on the contrary, that the head of the female has the fame colours with that of the male, but that its cafque is lefs elevated and more obtufe.

Roberts affirms, that it has not the cafque at all *.

Dampier and Labat maintain, that they never faw thofe red barbils and caruncles which border the noftrils in the male †.

Barrere tells us, that thefe parts are of a paler colour than in the male, and that the filky hairs of the *occiput* are thinner, fuch apparently as reprefented in Frifch's figure.

Laftly, the academicians found in fome individuals thefe filaments on the *occiput* rifing an inch, fo that they formed a kind of tuft behind the head.

It would be difficult, from all thefe varieties, to felect thofe that are fo deeply and fo permanently ftamped, as to conftitute diftinct races; and as we cannot doubt but that they are very recent, it will perhaps be fafeft to regard them as the effects produced by domeftication, change of climate, nature of the food, &c.; and without introducing them into the defcription, to mark the limits of the variations to which certain qualities of the Pintado are fubject, and to

* Voyage to the Cape de Verd iflands.

† New Voyage.—It is probable that the fhort and very bright red creft, mentioned by Father Charlevoix, is nothing but thefe caruncles.

endeavour,

endeavour, as much as poffible, to afcend to thofe caufes, of which the continued operation has at laft imprinted conftant characters, and formed diftinct fpecies.

In one circumftance, the Pintado bears a ftriking refemblance to the turkey; viz. it has no feathers on the head, nor on the arch of the neck. This has induced feveral ornithologifts, as Belon, Gefner, Aldrovandus, and Klein, to take the turkey for the Meleagris of the ancients. But not to mention the numberlefs points of difference between thefe two fpecies *, we need only refer to the proofs by which it was decided that the turkey was peculiar to America, and could never migrate into the ancient continent.

Briffon feems alfo to have miftaken, when, from a quotation of Kolben †, he inferted *Knor-baan*

---

* The Meleagris was, according to the ancients, as large as a good hen, and it had a flefhy tubercle on the head; its plumage was marked with white fpots like lentils, but larger; there were two barbils attached to the upper mandible, the tail was pendulous, the back round, there were membranes between the toes, and no fpurs at the feet: it delighted in marfhes, had no tendernefs for its young. Thefe characters are entirely different from thofe of the turkey, which, on the other hand, has many properties not to be found in the defcription of the Meleagris; particularly the bunch of hairs that hangs under the neck, and his manner of difplaying his tail, and of pacing around his female.

† " A bird which belongs properly to the Cape," fays this traveller, " is the *Knor-bahn*, or *Coq-knor*. It is the centinel of " the other birds; it informs them, when it fees a man approach, " by a fcream refembling the found of the word *crac*, and which

" it

haan in the lift of the names of the Pintado. I agree with Briffon, that the figure given by this traveller is only copied from that of the African Hen of Marcgrave; he muft alfo allow that it would be hard to admit a bird peculiar to the Cape of Good Hope to be the Pintado which is fpread through the whole of Africa, and lefs common at that promontory than in other parts of the country; ftill more difficult will it be to reconcile the fhort black bill, the crown of feathers, the red which is intermixed with the colours of the wings and of the body, and the quality which Kolben afcribes to his *Knor-haan*, that it lays only two eggs.

The plumage of the Pintado, though not decorated with rich and dazzling colours, is remarkably beautiful. It is of a bluifh-grey ground, fprinkled with confiderable regularity, with white roundifh fpeckles, refembling pearls. Hence fome of the moderns have beftowed on this bird the name of *Pearled Hen* *; and the ancients applied the epithets *varia* and *guttata* †. Such, at leaft, was the plumage in its native climate; but fince it has been carried into other

---

" it repeats very loud. It is as large as a common hen; its bill
" is fhort and black, like the feathers on its crown; the plumage
" of the wings and body is mixed with red, white, and cine-
" reous; the legs are yellow, and the wings fmall. It frequents
" folitary places, and builds its neft in the bufhes; it lays
" two eggs; its flefh is not much efteemed, though it is very
" good."

    * Frifch,   † Martial's Epigrams.

countries,

countries, it has affumed more of the white. Thus the Pintados at Jamaica and St. Domingo are white on the breaft; and Edwards mentions fome entirely white *. The whitenefs of the breaft, therefore, which Briffon confiders as the character of a variety, is only an alteration begun, in the natural colour, or rather it is the fhade between that colour and complete whitenefs.

The feathers on the middle part of the neck are very fhort near its arch, where they are entirely wanting. From that part they gradually lengthen unto the breaft, and there they are three inches long †.

Thefe feathers are of a downy texture from their root to near their middle, and this part is covered by the tips of the feathers in the preceding row, confifting of ftiff webs interwoven with each other.

It has fhort wings and a pendulous tail, like that of the partridge, which, joined to the arrangement of its feathers, makes it look as if it were hunch-backed (*Genus Gibberum*, PLIN.); but this appearance is falfe, and no veftige remains when the bird is plucked ‡.

The fize is nearly that of an ordinary hen, but the fhape is like that of the partridge; hence

---

* Gleanings, *Part Third.*

† Memoires pour fervir l'Hiftoire des Animaux, *Partie* II. p. 81.

‡ Lettres Edifiantes, *Recueil* xx.

it

it has been called the Newfoundland Partridge *.
But it is of a taller form, and its neck longer, and
more slender near the arch.

The barbils which rise from the upper mandible have no invariable form, being oval in
some, and square or triangular in others; they
are red in the female, and bluish in the male;
and, according to the academicians and Briſſon,
it is this circumſtance alone which diſtinguiſhes
the two ſexes. But other authors, as we have
already ſeen, have aſſigned different marks
drawn from the colours of the plumage †, of
the barbils ‡, the callous tubercle on the head §,
the caruncles of the noſtrils ‖, the ſize of the
body ¶, the ſilky threads of the *occiput*, &c. **;
whether theſe differences really reſult from the
ſex, or by a logical error, which is but too common, the accidental properties of the individual
have been regarded as ſexual.

Behind the barbils, we perceive on the ſides
of the head the very ſmall orifice of the ears,
which in moſt birds is concealed by feathers,
but in this is expoſed. But what is peculiar
to the Pintado is, a callous bump, or a kind
of caſque, which riſes on its head, and which
Belon improperly compares to the tubercle, or

---

* Belon.   † Caius *apud Geſnerum.*
‡ Columella, Friſch, Dampier, &c.
§ Aldrovandus, Roberts, Barrere, Dalechamp, &c.
‖ Barrere, Labat, Dampier, &c.
¶ Friſch.   ** Friſch, Barrere, &c.

rather

rather to the horn of the *giraffe* *. It refembles in fhape the reverfe of the ducal cap of the Doge of Venice, or this cap placed with its back towards the front. Its colour varies in different fubjects, from white to reddifh, paff-ing through the intermediate fhades of yellow and brown †. Its interior furface is like that of hard callous flefh, and it is covered with a dry wrinkled fkin, which extends over the *occiput*, and on the fides of the head, but is fur-rowed where the eyes are placed. Thofe natu-ralifts who deal in final caufes, have afferted, that this is a real helmet, beftowed on the Pin-tados as a defenfive armour, to protect them againft the attacks which they make on each other, becaufe they are quarrelfome birds, and have a ftrong bill and a delicate fkull.

The eyes are large and covered; the upper eye lid has long black hairs bent upwards, and the cryftalline lens is more convex at the ante-rior than at the pofterior furface.

Perrault affirms, that the bill is like that of the common hen; the Jefuit Margat makes it thrice as large, very hard, and pointed; the claws are alfo fharper, according to Labat. But

---

* It was on account of this tubercle that Linnæus termed the Pintado, in the fixth edition, *Hen with a* HORNY *top*; and in his tenth edition, *Pheafant with a* CALLOUS *top*.

† It is whitifh in Frifch; wax-coloured, according to Belon; brown, according to Marcgrave; tawny brown, according to Perrault; and reddifh in the *Planches Enluminées*.

all

all agree, both ancients and moderns, in faying
that the feet have no fpurs.

There is a remarkable difference which occurs between the ordinary hen and the Pintado;
that the inteftines of the latter are much fhorter
in proportion, not exceeding three feet, according to the academicians, exclufive of the
cæca, which are each fix inches, and widen as
they extend from their origin, and receive, like
the other inteftines, veffels from the mefentery.
The largeft of all is the *duodenum,* which is eight
lines in diameter. The gizzard is like that of
the common hen; and alfo contains numbers
of fmall pebbles, and fometimes even nothing
elfe; probably when the animal, dying of a lan-
guifhing diftemper, has paffed the clofe of its
life without eating at all.

The inner membrane of the gizzard is full of
wrinkles; it adheres loofely to the nervous coat,
and is of a fubftance analogous to horn.

The craw, when inflated, is about the fize of
a tennis ball; the duct, which joins it to the
gizzard, is of a harder and whiter fubftance
than what precedes the craw, and does not prefent near fo many diftinct veffels.

The *æfophagus* defcends along the neck, to
the right of the *trachea-arteria;* becaufe, no
doubt, the neck, which, as I have already faid, is
very long, bending oftener forwards than fide-
wife, the *æfophagus* preffed by the *trachea-
arteria,* whofe rings are entirely offeous, has
here,

here, as in moſt of the birds, been puſhed to that ſide where there is leaſt reſiſtance.

Theſe birds are ſubject to ſchirrous concretions in the liver, and even in the kidney. Some have been found without any gall-bladder; but in this caſe the hepatic branch was very thick. Others have occurred with only one teſticle; in general, it ſeems that the internal parts are no leſs liable to changes than the exterior and ſuperficial parts.

The heart is more pointed than common in birds[*]; the lungs are of the ordinary ſhape. It has however been obſerved in ſome ſubjects, that, on blowing into the *trachea-arteria* to inflate the lungs and air-cells, the *pericardium*, which appeared more than ordinarily flaccid, ſwelled with the lungs [†].

I ſhall add another anatomical remark, which has perhaps ſome connection with the habit of crying and the clamorous notes of the Pintado; it is, that the *trachea-arteria* receives in the cavity of the thorax two ſmall muſcular chords of an inch long, and two-thirds of a line broad, which are inſerted on each ſide [‡].

The Pintado is an exceedingly noiſy bird, and for this reaſon Brown has termed it *Gallus clamoſus* [§]. Its cry is ſharp, and by its con-

---

[*] Memoires pour ſervir a l'Hiſt. Nat. des Animaux.
[†] Hiſtoire de l'Academie des Sciences, tome i. p. 153.
[‡] Memoires pour ſervir a l'Hiſt. Nat. des Animaux.
[§] Natural Hiſtory of Jamaica.

tinuance,

tinuance, becomes fo troublefome, that, though the flefh is very delicate, and much fuperior to that of ordinary poultry, moft of the American planters have given over breeding it *.

The Greeks had a word appropriated to denote the fcreaming of the Pintado †. Ælian obferves, that the Meleagris utters a found refembling that of its name. Dr. Caius fays, that its cry is like that of the partridge, though not fo loud. Belon tells us, that it is analogous to the chirping of young chickens lately hatched; but at the fame time he pofitively affirms, that it is unlike that of ordinary hens. I cannot conceive why Aldrovandus and Salerne affert the contrary.

The Pintado is a lively, reftlefs, and turbulent bird, that diflikes to remain in the fame place, and contrives to become mafter of the poultry-yard. It can intimidate even the turkies; for, though much fmaller, it gains the afcendency over them by the mere dint of petulance. " The Pintado," fays Father Margat, " wheels " fometimes round, gives twenty ftrokes with " his bill, before thefe heavy birds are roufed " to defence." The Hens of Numidia feem to have the fame mode of fighting which the hiftorian Salluft imputes to the cavalry of that country. " Their charge is fudden and irre-

---

* Lettres Edifiantes, *Recueil* xx.

† Καγχαζειν, according to Pollux. *Gefner.*—That word fignifies alfo *to laugh loud.*

" gular;

" gular; if they meet with refiftance, they
" retreat, but in an inftant they renew the
" attack *." To this example we might add
many others, tending to prove the influence of
climate on the inftincts of the animals, as well
as on the national genius of the inhabitants.
The elephant joins to ftrength and induftry, a
flavifh difpofition ; the camel is laborious, patient, and fober; and, in thofe enervating
regions, even the dog forgets to bite.

Ælian relates, that in a certain ifland the
Meleagris is refpected by the birds of prey † ;
but I prefume that in every country of the
world, thefe would rather attack other fowls,
whofe bill is not fo ftrong, whofe head is not
protected by a cafque, and who are not fo well
acquainted with the art of defence.

The Pintado is one of thofe birds which feek,
by weltering in the duft, to rid themfelves of
infects.   They alfo fcrape the ground like common hens, and roam in numerous flocks.
Bodies of two or three hundred together are
fometimes feen in the Ifle of May; and the
inhabitants hunt them with a greyhound, and
without other weapons than fticks ‡.   But,
according to Belon, they run very faft, keeping their head elevated like the camelopard.

* Lettres Edifiantes, *Recueil* xx.
† *Hiftoria Animalium,* lib. v. 27,
‡ Dampier and Brue.

They

They perch at night to fleep, and fometimes
during the day, on the walls of inclofures, on
hedges, and even on the roofs of houfes, and on
trees. They are at great pains, Belon adds,
in providing their food; and, indeed, confidering the length of their inteftines, they muft
confume more than ordinary fowls, and be fubject to more frequent calls of hunger *.

It appears from the concurrence of the ancients † and moderns ‡, which is alfo corrobo-
rated by the femi-membranes which connect
the toes, that the Pintado is partly an aquatic
bird. Accordingly, thofe from Guinea, which
have recovered their liberty in St. Domingo,
and obey the impulfe of nature alone, prefer the
fwamps and moift fituations §.

If they be trained when young, they foon
become tame. Brue relates, that when he was
at the coaft of Senegal, he received, as a prefent from a princefs of that country, two Pintados, a male and a female, both of which were
fo familiar that they would come to eat on his

* De Seve obferved, in throwing fome bread to Pintados, that
when one of them happened to take a bit larger than it could immediately fwallow, it hurried away with it out of the reach of the
other fowls, and hid it in the dunghill, or in the earth, and fome-
time afterwards returned and ate it.

† Pliny, *Hiftoria Naturalis*, lib. xxxvii. 2. and Clitus of Miletus, in Athenæus.

‡ Gefner, Frifch,—Lettres Edifiantes.

§ Lettres Edifiantes.—" I entered," fays Adanfon, " a little
" thicket near a marfh, where flocks of Pintados were gathered."

plate; and that when they were at liberty to fly about the beach, they returned regularly to the ship, when the dinner or supper bell rung. Moore says, that they are as wild as the pheasants are in England *; but I suspect he never saw pheasants so tame as Brue's Pintados. And what proves that the Pintados are not very wild is, that they receive the food which is offered them the moment after they are caught.

The Pintado lays and hatches nearly like the ordinary hen; but its fecundity appears to be not the same in different climates, or at least that this is much greater in the domestic condition, where food is more abundant, than in the savage state, which affords but a scanty subsistence. I have been informed that it is wild in the Isle of France, and there lays ten or twelve eggs on the ground in the woods; whereas those that are domestic in St. Domingo, and seek the hedges and bushes to deposite their eggs, lay 100, or 150, provided that one be left constantly in the nest.

These eggs are smaller in proportion than those of an ordinary hen, and their shell is much harder. But there is a remarkable difference between those of the domestic Pintados and those of the wild sort; the latter are marked with small round spots like those on their plum-

* *Hist. Gen. des Voyages*, tome iii. p. 310.

age;

age; and this circumftance has not been overlooked by Ariftotle; but thofe of the former
are at firft of a pretty bright red, which afterwards fades, and at laft runs into the faint
colour of a dried rofe. If this fact be true, as
I have been affured by Fournier, who has raifed
many of thefe birds, we muft conclude that the
influence of domeftication penetrates here fo
deeply, as to change not only the colours of the
plumage, as we have already feen, but even
thofe of the matter which forms the fhell of the
eggs; and as this does not happen in other
fpecies, there is reafon to conclude that the nature of the Pintado is not fo fixed and invariable
as that of other birds.

Is the Pintado watchful or not of its brood?
This is a problem that has not yet been folved.
Belon replies without qualification in the affirma-
tive *. Frifch is of the fame opinion with regard to his great fpecies, which delights in dry
fituations, but affirms that the contrary is true
of the fmall fpecies, which prefers marfhes.
But moft authors impute to them a degree of
indifference for their offspring; the Jefuit Margat informs us, that at St. Domingo they are
not fuffered to cover their eggs, becaufe they
difcover fo little attachment, and fo often abandon their young †. The planters give their

* " They are very prolific, and careful in rearing their
" young." *Hiftory of Birds.*
† Lettres Edifiantes.

eggs

eggs to be hatched, he fays, under turkies or common hens.

I can find nothing with refpect to the time of incubation; but if we judge from the fize of the bird, and from our knowledge of other fpecies to which it is moft analogous, we may allow three weeks, more or lefs, according to the heat of the feafon or climate, and the affiduity of the fitter, &c.

In their firft infancy, the young Pintados have neither the barbles nor the cafque; they refemble the red partridges in their plumage, and the colour of their feet and bill, and it is difficult to diftinguifh the young males from the old females *; for in all thefe fpecies, the maturity of the females correfponds to the infancy of the males.

The young Pintados are very tender, and being natives of the burning climates of Africa, are with difficulty reared in our northern countries. According to Father Margat, they feed at St. Domingo, as well as the old ones, on millet. At the Ifle of May, they fubfift on the grafshoppers and worms, which they find themfelves by fcraping the ground with their nails †; and Frifch fays, that they live on all forts of grain and infects.

The Pintado cock breeds alfo with the common hen. But it is a kind of artificial union,

* I have this fact from Fournier, who was mentioned above.
† Dampier and Labat.

which

which requires attention to bring about. They muſt be bred together from their infancy; and the hybridous intercourſe gives birth to a baſtard progeny of an imperfect ſtructure, and diſavowed, as it were, by nature. Their eggs are deſtitute of the prolific power, and the race is extinguiſhed in the death of the individuals *.

The Pintados that are raiſed in our poultry-yards have an excellent flavour, in no reſpect inferior to that of partridges; but the wild or cheſnut ſort of St. Domingo have the moſt exquiſite reliſh, and exceed the delicacy of the pheaſant. The eggs of the Pintado too are a very agreeable food.

We have ſeen that the Pintado is of African origin; and hence all the names that have been beſtowed on it: hen of Africa, of Numidia, the foreign hen, that of Barbary, of Tunis, of Mauritania, of Lybia, of Guinea, of Egypt, of Pharaoh, and even of Jeruſalem. Some Mahometans called them Jeruſalem hens, and ſold them to the Chriſtians for whatever price they choſe to demand †; but theſe perceiving the fraud, retaliated on the good Muſulmen by offering them under the name of Mecca hens.

They are found in the iſles of France and Bourbon ‡, where they have been introduced

---

* Fournier.   † Longolius, *apud Geſnerum.*
‡ Aublet.

at

at a late period, but have since multiplied extremely *. They are known at Madagascar †
by the name of *acanques*, and at Congo by that
of *quetèle* ‡ ; they are very common in Guinea §,
on the Gold Coast, where they are kept tame
only in the district of Acra ‖ ; at Sierra-Leona ¶,
at Senegal **, in the island of Goree, in the
Cape de Verd islands ††, in Barbary, in Egypt,
in Arabia ‡‡, in Syria §§; we are not informed whether they occur in the island of Madeira, or in the Canaries. Gentil tells us, that
he saw Pintado hens at Java ‖‖ ; but it is uncertain if they were tame or wild : I should rather suppose that they were domestic, and carried from Africa to Asia, as they have been
transported from Europe to America. But as
these birds were accustomed to a hot climate, they
could not support the intense cold that reigns
on the frozen shores of the Baltic : and Linnæus never mentions them in his *Fauna Suecica*.
Klein seems to speak but from the report of

---

* Voyage Autour du Monde de la Barbinais le Gentil, tome xi.
p. 608.

† François Cauche, Relation de Madagascar.

‡ Marcgrave.   § Margat.

‖ Voyage de Barbot.   ¶ Marcgrave.

** Adanson's Voyage to Senegal.

†† Dampier's Voyage round the World.

‡‡ Strabo. *Lib.* xvi.

§§ " The most distant part of Syria breeds Pintados."
                                        DIODORUS SICULUS.

‖‖ Voyage round the World.

another

another perfon; and we are informed that at
the beginning of the prefent century they were
rare even in England *.

Varro fays, that in his time the African
hens, (it is thus he names the Pintados,) were
fold for a high price at Rome, on account of
their fcarcity †. They were much more common in Greece in the age of Paufanias; fince
this author pofitively afferts, that the meleagris, with the common goofe, was what perfons who were not in eafy circumftances, generally prefented at an offering in the folemn
myfteries of Ifis. But we muft not therefore
infer, that the Pintados were natives of Greece;
for, according to Athenæus, the Æolians were
the firft Greeks who were poffeffed of thefe birds.
Yet I conceive that fome trace of a regular
migration may be difcovered from the battles
that were annually fought with thefe birds in
Bœotia, on the tomb of Meleager; which are
mentioned both by the naturalifts and mythologifts ‡. Hence the name of Meleagris §, as that of Pintado ‖ has been be-

* Edwards's Gleanings. † De Re Ruftica, lib. iii. 9.
‡ Pliny, lib. x. 26.
§ According to the fable, the fifters of Meleager, having
gone diftracted through exceffive grief at their brother's death,
were turned into thefe birds, which ftill bear the tears fprinkled on
their plumage.
‖ *Peintade*, from *peindre*, which, in French, fignifies to
paint.

M 4   ftowed,

ftowed, on account of the beautiful diftribution of the colours with which their plumage
is painted.   [A]

[A] Specific charaĉter of the Pintado, *Numidia Meleagris*:
" Has a double caruncle at the chaps, no fold at the throat.''
Mr. Pennant makes it appear that the Pintados had been early
introduced into Britain ; at leaft prior to the year 1277.   But they
feem to have been much negleĉted, on account of the difficulty of
rearing them ; for they occur not in our ancient bills of fare.

M

THE WOOD GROUS .

The WOOD GROUS.

Le Tetras, ou *le Grande Coq de Bruyere* *, Buff.
Tetrao-Urogallus † Linn. Gmel, &c. &c.
Urogallus Major, Briff. Klein, and Gerini.
The Capcalze, Sibbald, Scot. Illuft.
The Cock of the Wood, or Mountain, Ray, Will. and Alb.

IF we were to judge of things by their names
only, we fhould take this bird for a wild
cock or a pheafant; for in many countries, particularly in Italy, it is called Wild Cock ‡,
gallo alpeftre, felvatico §. In other places, it is
termed the *Noify Pheafant,* and the *Wild Pheafant.* But it differs from the pheafant in its
tail, which is of another fhape, and only half
the length; in the number of great feathers that
compofe it; in the extent of its wings compared

---

* *i. e.* The Great Heath-Cock.

† In Greek, τάριξ, which was probably formed from τέίριγως,
the participle middle of the verb τριζω, to make a creeking noife,
alluding to the whirring cry of the Grous. The word τέίραων,
a-kin to the former, feems to have been in ufe, and hence the
Latin *Tetrao. Auer,* in old German, fignifies *fhy* or *wild,* and
the Grous was therefore termed *Auer-hahn,* or Wild Hen, which
was latinized into *Uro-gallus.* In Italian it is called *Gallo Cedrone,*
or the Cedar Cock. In Polifh it is named *Glufzec;* in Swedifh,
Kjaeder; and in Norwegian, *Lieure.*

‡ Albin defcribes the male and female under- the name of the
Black Cock and Hen of the Mufcovy mountains.

§ *i. e.* Mountain Cock, Wood Cock.

with

with its other dimenfions; and in the form of its
feet, which are rough and without fpurs, &c.
Befides, though both thefe fpecies of birds delight in forefts, they are feldom found in the
fame fpots; the pheafant, which fhuns cold,
fixes its refidence in the woods that grow in the
plains; while the Grous prefers the chill expofure of the woods which crown the fummits
of lofty mountains. Hence the names of *Cock
of the Mountain,* and *Cock of the Wood.*

Thofe who, with Gefner, and fome others,
would confider it as the original cock, can indeed found their conjecture on fome analogies;
the general fhape of its body; the particular
configuration of its bill; the red projecting fkin
above the eyes, the fingular nature of its feathers, which are moftly double, and rife in pairs
from the fame root, a property which, according to Belon, is peculiar to the ordinary
cock; and laftly, they have the fame common
habits, one male fupplying feveral females, and
thefe not building any nefts, but fitting on their
eggs with much affiduity, and fhowing a ftrong
affection to their young after they are hatched.
But if we confider that the Grous has no membranes under its bill, and no fpurs on its feet;
that its feet are clothed with plumage, and its
toes are edged with a kind of indenting; that
there are two quills more in the tail; that this
tail is not divided into two planes as in the
ordinary cock, but can be difplayed like a fan

as

as in the turkey; that its bulk is quadruple that
of the ordinary cock; that it is fond of cold
countries, while the domeftic fowls thrive beft
in temperate climates ; that no inftance was
ever adduced of the intermixture of the breeds ;
and that their eggs are of a different colour : If
to all thefe we add the proofs already given,
that the ordinary cock is a native of the genial
regions of Afia, where travellers have hardly
ever feen the Grous; we certainly cannot admit
that thefe are the primitive ftock, and we muft
impute it to an error occafioned, like many
others, by the deceitful glofs of names.

Ariftotle merely mentions a bird which he
terms *tetrix*, and which the Athenians called
ourax, (υραξ); it is a bird, he fays, which
does not neftle on trees or on the ground, but
among low creeping plants *. A little afterwards,
he adds, that the *tetrix* does not make any neft,
but drops its eggs on the ground like all the heavy
birds, and covers them with ftiff herbs. This fhort
defcription manifeftly applies to the Grous, the
female of which conftructs no neft, but drops
her eggs on mofs, and when obliged to leave
them, covers them carefully with leaves. Befides, the Latin word *tetrao*, which Pliny employs
to fignify the Grous, has an evident analogy to
the Greek *tetrix*, not to mention the refemblance
which the Athenian *ourax* bears to the com-

* Εν τοις χαμαιζιλοις φυλοις, Lib. vi. 1.

pound

pound term *ourh-hahn* beftowed by the Germans, a coincidence which cannot with propriety be afcribed to chance.

But there is a circumftance which feems to
fhed fome doubts on the identity of thefe birds.
Pliny, defcribing his *tetrao* at fome length, never
takes notice of what Ariftotle had faid of the
tetrix, which it is likely he would have done, if
he had conceived thefe to be the fame; unlefs
the flight mention made by Ariftotle had efcaped
the Roman naturalift.

With regard to the great *tetrax* of which Athenæus fpeaks *, it is certainly not our Grous,
fince it has flefhy barbles like thofe of the cock,
rifing near the ears and defcending below the
bill; a character quite foreign to the Grous, and
which applies much better to the Meleagris or
Numidian hen, which is our Pintado.

The little *tetrax* mentioned by the fame
author, is, according to him, an exceeding
fmall bird; and this excludes all comparifon
with our Wood Grous, which is one of the firft
magnitude.

In refpect to the *tetrax* of the poet Nemefianus, who dwells on its ftupidity, Gefner
confiders it as a fpecies of buftard. But I difcover a difcriminating mark of refemblance to
the meleagris in the colours of its plumage; the

* Lib. ix.

4   ground

ground is of an afh-gray, fprinkled with fpots in the fhape of drops * ; a circumftance which has given the pintado the name of *Gallina Guttata* † .

But whatever be the force of thefe conjectures, it appears inconteftably proved, that the two fpecies of the *tetrao* of Pliny are really thofe of our Grous: the fine fhining black of their plumage; the flame-colour of their eyebrows; their refidence in cold mountainous countries ‡ ; the delicacy of their flefh; thefe are properties that belong both to the Wood and Black Grous. We can even diftinguifh in Pliny's defcription, the traces of a peculiarity that has been remarked by few moderns; *Moriuntur contumaciâ*, fays this author, *fpiritu revocato* §. This refers to a curious obfervation which Frifch has inferted in his hiftory of this bird. That naturalift, not being able to find

* *Fragments of Books on Bird-catching* ; afcribed by fome to the Poet Nemefianus, who lived in the third century of the Chriftian æra.

† *Et pidla perdix, Numidicæque guttatæ*. " And the painted " partridge, and the fpeckled Numidian hens." MARTIAL. This is exactly the plumage of the two hens belonging to the Duke of Ferrara, of which Gefner fays, in his account of the Pintado: " That they were entirely of a cinereous colour, with a " whitifh caft, and with black and round fpots."

‡ " A gloffy jet black becomes the *tetraones*, and a fcarlet on the eye-brows.—They inhabit the Alps, and the region of the North."—PLINY, lib. x. 22. The *tetrao* feen by Belon on the lofty mountains of Crete, correfponds well to Pliny's defcription.

§ *i. e.* They die through obftinacy, recalling their breath.

the

the tongue in a dead Grous, opened the gizzard, and difcovered that it retreated there with all its ligaments; and this muft commonly happen, fince it is the general opinion of fportfmen that the Grous has no tongue. The fame, perhaps, might be faid of the Black Eagle mentioned by Pliny, and the Brazil bird of which Scaliger fpeaks, which was reckoned to have no tongue. This opinion might take its rife with credulous travellers, or unobfervant hunters, who never viewed this bird except when expiring, or after death, and no perfon infpecting their gizzard.

The other fpecies of *tetrao*, which Pliny defcribes at the fame place, is much larger; fince it exceeds the buftard, and even the vulture, which it refembles in plumage, and in point of fize is inferior to the oftrich alone: befides, it is fo unwieldy a bird, that it can be caught by the hand *. Belon afferts, that this fpecies of *tetrao* is unknown to the moderns, who, according to him, have never feen any Wood Grous larger, or even fo large as the buftard; and there is room to doubt, whether the bird mentioned in this paffage of Pliny by the names *Otis* and *Avis tarda*, was really our buftard, whofe flefh has an excellent flavour, while the *avis tarda* of Pliny was very unpa-

---

* This is literally true of the Little Grous, as we fhall find in the following article.

latable.

latable. But we muft not, on this account, infer with Belon, that the great *tetrao* was no other than the *avis tarda*; fince the Roman naturalift names both the *tetrao* and the *avis tarda* in the fame paffage, and compares them together as birds of different fpecies.

After a mature confideration of the fubject, I fhould rather conclude: 1. That the firft *tetrao* of which Pliny fpeaks, is the fmall fpecies of Grous, to which what is here faid more directly refers; 2. That his great *tetrao* is our Wood Grous, which, without exaggeration, exceeds the bulk of the buftard. I myfelf weighed a large buftard, whofe extreme length was three feet three inches, and the extent of whofe wings was fix feet and an half, and found it twelve pounds; but it is well known, and we fhall afterwards have occafion to take notice of it, that fome of the Wood Grous weigh more.

The Wood Grous has near four feet of alar extent. Its weight is generally twelve or fifteen pounds: Aldrovandus affirms, that he has feen fome that were twenty-three pounds; but thefe were Bologna pounds, which contain each only ten ounces, and therefore twenty-three are not quite equal to fifteen pounds of fixteen ounces. The Black Cock of the Mountains of Mufcovy, defcribed by Albin, and which is really the Wood Grous, weighed ten pounds without the feathers or entrails; and the
fame

fame author informs us, that the *lieures* of Norway, which is really the fame bird, is as large as a buftard.

This bird fcrapes the ground, like all the frugivorous tribe. Its bill is ftrong and fharp *; the tongue is pointed, and lodged in a proportional concavity in the palate. The feet are alfo firm, and clothed before with plumage; the craw is extremely wide, but, in other refpects, both it and the gizzard are conftructed as in the domeftic cock: the coat of the gizzard has a velvet foftnefs where the mufcles are attached.

The Wood Grous feeds on the leaves or tops of the pine, of the juniper, of the cedar, of the willow, of the white poplar, of the hazel, of the myrtle, of the bramble; on thiftles, fir-cones, the leaves and flowers of buck-wheat; on chichling vetch, millfoil, dandelion, trefoil, the vetch, and the choke-weed; efpecially when thefe plants are young and tender. When the feed begins to be formed, they leave the flowers, and only eat the leaves. They feed too, efpecially in their firft year, on black-berries, beech-maft, and ants eggs. On the other hand, it has been obferved, that many

---

* I know not what Longolius means, when he fays that this bird has traces of barbils. Is there a kind of large grous which have barbils, as there is among the fmall grous; or does he allude to a certain difpofition of the feathers reprefenting, imperfectly, barbils, as he has done at the article of the Hazel Grous?

plants

plants prove poifonous to this bird; among
others, lovage, celandine, wall-wort, lily of the
valley, wheat, nettles, &c. *

On opening the gizzard of the Wood Grous,
fmall pebbles have been found, fimilar to thofe
in common poultry; a certain proof that they
do not confine themfelves to the leaves and flowers which they pluck from the trees, but alfo
feed on grains which they feek by fcraping the
ground. When they eat too many juniper-
berries, their flefh, which otherwife is excellent,
contracts an unpleafant tafte; and according to
the remark of Pliny, it lofes its delicate flavour,
if kept in cages or coops, where it is fometimes
fed for curiofity †.

The female differs from the male only in its
fize and plumage, being fmaller and not fo black;
befides, it excels the male in the beauty and variety
of its colours; a circumftance which is uncommon
in birds, and even in other animals. From not attending to this fact, Gefner has made the female
another fpecies of Grous, by the name of *grygallus
major*, formed from the German term *grugel-
hahn*; for the fame reafon, he has made the female of the Black Grous another fpecies, which
he calls *grygallus minor*. Yet he pretends that
he did not fix thefe fpecies till he had carefully
examined all the individuals, except the *grygallus minor*, and was confident that he could

* Journal Economique, *Mai* 1765. † Lib. x. 22.

perceive

perceive characteriftical differences. On the other hand, Schwenckfeld, whofe refidence was among mountains, and who obferved the *gry-gallus* often and carefully, affures us, that it was the female Grous; but it muft be admitted that in this fpecies, and perhaps in many others, the plumage is fubject to great diverfity, arifing from the age, the fex, the climate, and other cir-cumftances. The one which we have caufed to be engraved is fomewhat crefted. Briffon takes no notice of a creft in his defcription; and of the two figures given by Aldrovandus, the one is crefted, the other not. Some pretend that the Grous, when young, has much white in its plumage *, which diminifhes, as the bird grows old, and fo regularly as to ferve as a mark for diftinguifhing the age. It would even appear that the number of quills in the tail is not con-ftant; for Linnæus makes it eighteen in his *Fauna Suecica*, and Briffon only fixteen in his *Ornithology;* and what is more extraordinary, Schwenckfeld, who faw and examined many of thefe birds, afferts that, both in the large and in the fmall fpecies, the females have eighteen tail-feathers, and the male only twelve. It therefore follows, that every fyftem which af-fumes, for its fpecific characters, differences fo variable as are the colours and even the number of the feathers, will be liable to the great incon-

---

* When the bird difplays its tail, the white forms a circle around it.

venience

venience of multiplying the species, (I should rather say, nominal species, or more properly new terms,) of oppressing the memory of beginners, and of giving them false ideas of things, and consequently that such a plan increases the difficulties in the study of nature.

It is false, what Encelius relates, that the male Grous sitting on a tree, calls the females to him with loud cries, emits *femen* from his bill, which they swallow, and again discharge, and thus become impregnated. Nor is it true, that the part of the *femen* which is not gathered up by the females, forms serpents, precious stones, and pearls. It is mortifying to our pride to see the human mind inflicted with grovelling errors, or hurried into such extravagant follies. The Grous couple like other birds; nor was Encelius unacquainted with the fact; but he insists that the embrace is mere dalliance, and that the deglutition of the *femen* is essential to propagation !

The male Grous begins to be in season about the first of February; the fiery appetite is most intense towards the latter end of March, and continues till the leaves are expanded. During that period of love, each cock fixes his residence in a certain quarter, out of which he never removes. In the morning and evening he is observed walking backwards and forwards on the trunk of a large pine or other tree, his tail displayed, his wings trailing, his neck projecting, his head ruffled, and assuming all sorts of un-

common

common poftures; with fuch force is he impelled by the burning defires! He has a certain
note with which he calls his females, who run
under the tree where he lodges, from which he
foon defcends to tafte the joys of love. This
fingular cry, which is very loud, and can be
heard at a great diftance, is perhaps the reafon
of the name which has been applied, of *noify
pheafant*. It begins with a kind of explofion,
and expires in a fharp fhrill note, refembling the
found produced by whetting a fcythe. This
noife vanifhes and returns alternately, and after
being repeated feveral times in the courfe of half
an hour, it ends in explofion like the firft *.

The Wood Grous, which at other times is
very fhy, can eafily be furprifed in the feafon of
love, efpecially when it is occupied with its call:
it is then ftunned with its own noife, or, if we
chufe, fo intoxicated, that it is neither fcared by
the fight of man, nor roufed by the report of a
fowling-piece. It fees nothing, it hears nothing, it is diffolved into extacy †; hence it has
been faid and even written, that the Grous is
deaf and blind. But almoft all animals, not excepting man, are, in fimilar fituations, abforbed
in delight: all feel, in a certain degree, the rap-

* Journal Economique, *April* 1753.
† " It is fo overgrown that it may be caught motionlefs on the
ground." PLINY. What that naturalift imputes to its bulk,
may be referred with greater probability to heat, and intoxication
of its paffion.

tures

tures of maddening joys. But probably the Wood Grous is more under the dominion of luft; for in Germany, the term *auerhabn* is beftowed on the lover, who neglects every other concern, and devotes himfelf entirely to the object of his paffion*, and even applied to every perfon who difcovers a ftupid infenfibility to his moft important interefts.

It will be readily conceived that the feafon when the Grous is wholly occupied by the amorous paffions, is the proper time for fetting gins, or for hunting it. When I come to treat of the Small Black Grous, I fhall defcribe more particularly the precautions obferved in this fport; I fhall here only obferve, that people are attentive in extirpating the old cocks, becaufe thefe appropriate an extenfive tract, and fuffer no rivals to enter the region of their pleafures; and thus many females are deprived of the male influence, and produce addle eggs.

Some bird-catchers pretend, that before the Wood Grous couple, they provide a clean even fpot †. That fuch may occur I have no doubt, but I fufpect that the Grous fhow no forefight in choice. It is much more natural to fuppofe that thefe fpots have been the habitual refort of the hen and her young, and that after two or three months they become more trodden and flat than the reft of the ground.

* Frifch. † Gefner.

N 3   The

The leaft number of eggs which the female Wood Grous commonly lays, is five or fix ; the greateft number, eight or nine. Schwenckfeld afferts, that their firft hatch is eight, and the fubfequent ones amount to twelve, fifteen, or even fixteen *. Thefe eggs are white, fpotted with yellow ; and, according to the fame author, they are larger than thofe of common hens. The female drops them in a dry fpot on the mofs, where it hatches them alone, without the affiftance of the male †. When it is obliged to leave the eggs, it carefully ftrews them with leaves ; and though it inherits a favage inftinct, the love of progeny feems to blunt the fenfe of immediate danger, and it continues to fit after we have approached it, and can hardly be forced to forfake its eggs.

As foon as the young are hatched, they run nimbly, and even before the fhell is completely detached. The mother leads them in the moft careful and attentive manner ; fhe goes with them into the woods, where fhe feeds them with ants eggs, black-berries, &c. They continue united through the reft of the year, till the return of the feafon of love, infpiring them with new appetites and inclinations, difperfes the family ; the males are the wideft feparated, never affociating

* This gradation is conformable to the general obfervation of Ariftotle ; I fufpect only that the number is over-rated.

† I have fomewhere read that the time of incubation is twenty-eight days, which is probable, confidering the bulk of the bird.

with

with their own fex, and feldom mixing with the females but to fatiate their luft.

The Wood Grous delights, as we have already obferved, in lofty mountains. But this is the cafe only in the milder latitudes; for in countries that are intenfely cold, as Hudfon's Bay, they prefer the plains and fheltered places; and in fuch fituations, they enjoy, in thofe inclement regions, perhaps the fame temperature as on the moft elevated fummits in the genial climes *. They inhabit the Alps, the Pyrenees, the mountains of Auvergne, of Savoy, of Switzerland, of Weftphalia, of Swabia, of Mufcovy, of Scotland, thofe of Greece and Italy, thofe of Norway, and even thofe in northern tracts of the continent of America. It is fuppofed that the breed is extinct in Ireland, where however they once refided †.

It is faid that birds of prey are very deftructive to them; either becaufe they direct their affaults when the Wood Grous is intoxicated with love, or growing fond of the fuperior delicacy of their flefh, they felect them for their prey. [A]

* Hift. Gen. des Voyages, tome xiv.
† Britifh Zoology.

[A] Specific character of the Wood Grous, *Tetrao-Urogallus:* " Its tail is rounded, its axillary feathers white." Mr. Pennant, whofe authority on this fubject is unqueftionable, affures us, that this bird is not found in America. It is now almoft extinct in Scotland, being found only in the forefts north of Lochnefs.

The BLACK GROUS.

Le Petit Tetras, ou Coq de Bruyere à Queue Fourchue *, Buff.
Tetrao-Tetrix, Linn. Gmel. &c. &c.
Urogallus Minor, Briff.
Gallus Scoticus Sylveſtris, Aldrov.
The Black Cock, Sibbald. Scot. Illuſt.
The Heath-Cock, Black-Game, or Grous †, Will.

SOME authors, as Rzacynſki, have miſtaken this bird for the *tetrax* of the poet Neme-
ſianus. This overſight aroſe undoubtedly from not attending to what Nemeſianus himſelf had mentioned, that it was of the bulk of a gooſe or a crane ‡; ſome other obſervers inform us, that the Black Grous is ſcarcely larger than an or-dinary cock, but only longer ſhaped; and the female, according to Ray, is ſmaller than a com-mon hen.

Turner, ſpeaking of his Mooriſh hen, ſo call-ed, he ſays, not on account of its plumage, which

* i. e. The forked-tail Heath-cock.
† This bird has alſo been termed improperly a cock or phea-
ſant : *Little Wild Cock (Petit Coq Savage)*; *Heath Cock (Coq de
Bruyere)*; *Birch-Cock, &c. (Coq de Bouleau)*; *Black Pheaſant
(Faiſan Noir)*; *Mountain Pheaſant (Faiſant de Montagne)*. In
German, *Birkhan (Birch-Hen)*; in Swediſh, *Orre*; the ſame
with the old German *Eure,* mentioned before; in Norwegian,
Orrfugl (the *Eure-bird*).

‡ Tarpeiæ eſt euſtos arcis non corpore major,
Nec qui te volucres docuit, Palamede, figuras.

reſembles

THE BLACK GROUS.

refembles that of the partridge, but on account
of the colour of the male, which is black, afcribes
to it a red flefh comb, with two barbils of the
fame colour and fubftance *. In this affertion
Willughby infifts that he was miftaken. But it
is difficult to conceive that Turner could fall into
an error with refpect to this bird, which inhabits
his own country, and concerning a character
that would be fo eafily noticed. On the other
hand, admitting what Turner fays, I fhould refer his Moorifh hen to another fpecies; or, if we
chufe, to another fort of the Black Grous, analogous to the firft in its general ftructure and habits,
but diftinguifhed by its undivided tail and its
flefh barbils; and what confirms me in this
opinion is, that I find in Gefner, a bird by the
name of *gallus fylveftris*, which has thefe properties; fo that we may confider it as an individual of the fame fpecies with Turner's Moorifh
hen; efpecially, as in this fpecies, the male is
called the *black cock* in Scotland, (whence Gefner
received his figure,) and the female *grey hen*; a
circumftance which marks diftinctly the difference of the plumage of the two fexes in this
fpecies of the Grous.

The Black Grous weighs three or four pounds.
It bears a great refemblance to the Wood Grous;
it has red eye-lids, rough feet without fpurs,
indented toes, white fpot on the wing, &c. But
it is diftinguifhed by two obvious characters; it

* See Gefner.

is much fmaller, and its tail is forked, the outer
feathers being longer, the middle ones bent backwards., Befides, the male of the fmall fpecies is
of a deeper and more diftinct black; the red glandulous fkin above the eyes is broader, but fubject to fome variations in the fame individuals at
different times, as we fhall find in the fequel.

The female is only two thirds of the fize of
the male*. Its tail is lefs forked, and the colours of its plumage are fo different, that Gefner
was induced to refer it to a diftinct fpecies, by
the name of *grygallus minor*. This change in
the colours of the plumage does not take place
till after a certain age; the young males at firft
refemble their mother, and preferve the fame
appearance till the end of the autumn. Towards
the clofe of that feafon, and during the winter,
the plumage gradually acquires a deeper colour,
till it becomes a bluifh-black, which is permanent
thenceforth, except the flight changes which I
fhall mention. 1. The blue increafes fomewhat
with age: 2. At the end of three years, and not
fooner, a white fpot appears under the bill:
3. When they are very old, another fpot of a
variegated black fpreads under the tail, where the
feathers are all white † Charleton, and fome
others add, that the number of white fpecks on
the tail diminifhes regularly with the age of the
bird, fo as to ferve for a mark to difcover it.

---

* Britifh Zoology.   † Acts of Breflau, Nov. 1725.

The

The naturalists who unanimously reckon twenty-six quills in the wing of the Black Grous, do not agree so well with respect to the number of quills in the tail: Schwenckfeld allows eighteen to the female, and only twelve to the male. Willoughby, Albin, and Brisson, bestow sixteen on either sex. The two males preserved in the Royal Cabinet have each eighteen; viz. seven large ones on each side, and four in the middle much shorter. Must we ascribe these differences to a real variation in the number of quill-feathers; or are we to impute them to the inaccuracy or inattention of the observers?—The wings of the Black Grous are short, and hence its flight is laborious, nor is it ever seen to rise high, or to pursue a distant course.

In both sexes the orifice of the ears is wide, the toes are connected by a membrane as high as the first articulation, and edged with indenting; their flesh is white, and of easy digestion; the tongue soft, beset with small points, and not parted; under the tongue is a glandulous substance; in the palate, a cavity corresponding exactly to the dimensions of the tongue; the craw is very large, the intestinal tube fifty-one inches long, and the appendices or *cæca* twenty-four; these fluted with six *striæ* *.

The difference between the male and female is not confined to the surface; it penetrates even

---

* Willughby and Schwenckfeld.

to the interior organization. Dr. Waygand obferves, that the bone of the *fternum* in the males,
being held to the light, appears interwoven with
a prodigious number of fmall ramifications of a
red colour, which meander in every fhape and
in all directions, and form a curious and fingular
web; but that in the females the fame bone
has nothing analogous to thefe ramifications;
it is befides very fmall, and of a whitifh
colour *.

This bird flies often in flocks, and perches on trees
much like the pheafant †. It cafts its feathers
in the fummer, and then conceals itfelf in luxu-
riant heath, or feeks for lodgment among fens ‡.
It feeds chiefly on the leaves and buds of the
birch, or on the berries that are the fpontaneous
production of Alpine tracts. Hence the French
name *coq de bruyere*, or *heath-cock*; and the German of *birkhan* or *birch-hen*. It alfo eats the catkins of the hazel, wheat, and other grains; in
autumn, it has recourfe to the acorns, bramble-
berries, alder-buds, pine-cones, bilberries, and
the berries of the fpindle tree; in winter, it retires to the extenfive forefts, and fubfifts on juniper berries, or fearches under the fnow for
the cranberries §. Sometimes it lives two or
three months, in the rigour of winter, without

* Acts of Breflaw, as quoted above.
† Britifh Zoology.   ‡ Acts of Breflaw.
§ Schwenckfeld.—Rzacynfki.—Willughby, and the Britifh
Zoology.

any

any food *; for in Norway it is faid to pafs the
inclement feafon, torpid and motionlefs beneath
the fnow; in the fame manner, as in the milder
climates the bats, the dormice, the *lerots*, the
fhrew-mice, and the marmots, (if the fact be
true) fuffer a temporary fufpenfion of the active
powers †.

Thefe birds are found in the mountainous
tracts of the North of England and Scotland;
in Norway, and the boreal provinces of Sweden;
in the neighbourhood of Cologne; in the Swifs
Alps; in Bugey, where, according to Hubert,
they are called *grianots*; in Podolia; in Lithuania; in Samogitia; and particularly in Vol-
hinia, and in the Ukraine, which includes the
Palatinates of Kiovia and Breflaw, where a Polifh noble, as Rzacynzfki fays, caught in one
day, near the village of Kufmince, one hundred

* The author of the Britifh Zoology remarks, that the white
partridges which winter in the fnow, have their legs better clothed
with feathers, than the two fpecies of Grous which find fhelter in
the thick forefts. But if the Grous fleep beneath the fnow, what
becomes of this final caufe, or rather what becomes of all that fuperficial fort of reafoning when examined by the light of philofo-
phy?

† This puts me in mind of what is related in the book *De Mi-
rabilibus*, afcribed to Ariftotle, that certain birds in the kingdom of
Pontus lay during the winter in fuch a ftate of torpor, that they
might be plucked, and even ftuck upon the fpit, without fhewing
any feeling, and were not roufed from their lethargy till they
began to be roafted. If we ftrip this tale of the marvellous, it
alludes to the fame fort of torpor with that of the Grous and
Marmots, while the functions of the external fenfes are fufpended
for want of heat.

5

and

and thirty brace, in a fingle drawing of the net.—
We fhall afterwards fee the mode of catching them
which is practifed in Courland. Thefe birds can
hardly be reconciled to a different climate, or to
their domeftic ftate; almoft all thofe which
Marfhal Saxe got from Sweden for his *menagerie*
at Chambor, died of melancholy, without leaving pofterity *.

The Black Grous comes in feafon about the
time when the willows begin to fhoot, that is
towards the end of the winter; the fportfmen
readily difcover it by the humidity of their excrements †. It is then that the males are obferved to affemble by day-break, to the number
of a hundred or more, in fome place which is
elevated, folitary, furrounded with marfhes, or
covered with heath, and this is the field of continual contention; they fight bitterly with each
other, till the vanquifhed are driven to flight.
The victors then feat themfelves on the trunk
of a tree, or on a rifing fpot of ground, their eyes
flafhing fire, their eye-brows fwelled, their feathers briftled, their tail expanded like a fan;
beating their wings and frifking with wild defire ‡, they invite their females by a call, which
may be heard at half-a-mile's diftance; the natural note which refembles the found of the
German word *frau* §, rifes at this time one
third, and is joined with another fingular cry,

* Salerne.   † Acts of Breflaw, Nov. 1725.
‡ Frifch.—Britifh Zoology.   § Salerne.

or kind of noify rattling of the gizzard *. The females in the neighbourhood anfwer to the voice of the males, by a cry peculiar to them, flock around their mates, and in the following days refort to the precife fpot. According to Dr. Waygand, each cock has two or three hens, to which he is more particularly attached.

When the females are impregnated, they retire to lay their eggs in copfes, which are thick and tall. They drop them on the ground, and, like all the large birds, are at little pains in conftructing a neft. They lay fix or feven eggs according to fome †; from twelve to fixteen ‡; and even from twelve to twenty, according to others §; thefe are not fo large as thofe of the domeftic hens, but are fomewhat longer fhaped. Linnæus afferts, that the female Black Grous lofes its delicate flavour in the time of incubation. Schwenckfeld feems to infinuate that their feafon for laying is now deranged, fince they have been molefted by the fportfmen, and fcared by the reports of the fowling-piece; and to the fame caufes he afcribes the extinction in Germany of many other beautiful fpecies of birds.

As foon as the chickens are twelve or fifteen days old, they flap their wings, and effay to fly; but it is five or fix weeks before they are able to rife from the ground, and then they perch on the

---

* Frifch.   † Britifh Zoology.   ‡ Schwenckfeld.
§ Acts of Breflaw.

trees with their mothers.   This is the time to
decoy them with a call *, to catch them in the
net, or to fhoot them.   The mother miftaking
this call for the chirping of her ftrayed young,
runs to the place, and invites them by a particular cry, which fhe often repeats, like the domeftic
fowls in the fame circumftances ; fhe thus collects the whole covey, and all become devoted to
the mercy of the fportfmen.

As they grow bigger, their plumage gradually
affumes a black caft, and then they are not fo
eafily decoyed ; but when they have attained
half their growth, the falcon is flown at them ;
and the proper time is about the clofe of autumn,
when the trees have fhed their leaves.   In that
feafon, the males felect fome fpot, whither they
repair every morning at fun-rife, and by a certain cry (efpecially when it is likely to be froft,
or fine weather), they invite all other birds of the
fame fpecies, of every age and of either fex.
When affembled, they fly in flocks to the bufhes ;
or if there is no fnow on the ground, they difperfe over the ftubble fields, where barley, oats,
or other fuch grain has been reaped.   Then it is
that birds of prey trained on purpofe afford excellent fport.

Another method of catching this game is practifed in Courland, Livonia, and Lithuania. They
ufe a ftuffed grous, or an artificial bird made of

* This call is made of a bone of the Gofhawk, which is filled
with wax, and proper holes bored in it.   BRESLAW's *Acts.*

cloth

cloth of the proper colour, and ftuffed with hay
or tow, and this is termed in thofe countries,
balvane. They faften this *balvane* to the end of
a ftick, and place it on fome birch-tree near the
fcene of their amours : the time for this fport is
in the month of April. The birds gather round
the *balvane*, and fight with each other in play ;
at laft they engage in earneft, and are fo much
occupied in the violence of their contentions,
that the fportfman, who is concealed near the
fpot in his hut, furprifes them, and catches them,
without being obliged to aim a fingle blow.
Thofe caught in this way, he tames in the fpace
of five or fix days ; fo that they will come to eat
out of his hand *. On the following year in the
fpring, they make ufe of thofe tame birds, inftead of *balvanes*, to decoy the wild Black
Grous, which fall upon them, and fight with
fuch fury as not to be fcared by the report of a
fowling-piece. Each morning they repair by
day-break to the common rendezvous, and remain there till fun-rife, when they fly away and
difperfe through the forefts and heaths in fearch
of food. About three o'clock in the afternoon,
they return to the fame fpot. and continue there
till late in the evening. This is their regular
courfe of life, efpecially in fair weather, during
the feafon of love, which lafts three or four

---

* In this refpeft the Little Grous differs widely from the Great
Grous, which, far from fubmitting to domeftication, conftantly re-
jects what is offered it to eat.

weeks ;

weeks; but when it is rainy or cold, they are rather more retired.

The young Black Grous have also their favourite fpot of refort, where they affemble in flocks of forty or fifty at a time, and devote themfelves to nearly the fame amufements or occupations; their voice however is hoarfer, and broken; and they do not leap with equal agility. Their meeting lafts only eight days, after which they join the old ones.

When the feafon of love is over, and confequently their affemblies lefs regular, new ftratagems muft be employed to decoy them near the hut where is the *balvane*. Several fportfmen on horfeback enclofe a circuit of variable extent, having the hut for its centre, and cracking their whips, they drive the Grous from bufh to bufh, and fo gradually contraƈt the bounds, and, by means of a whiftle, they inform the perfon who manages the *balvane* of their approach. The Grous, when they fhift from one bufh to another, diftinguifh accurately thofe branches which are able to fupport them, not even excepting the vertical fhoots, which bend with their weight into an horizontal fituation; after alighting they liften attentively, ftretching out their neck to learn if they are in a place of fafety, and as foon as they have allayed their fears, they begin to pluck the tender buds. The dexterous fportfman then feizes the opportunity of placing his *balvane* on the neighbouring twigs, and

3

fixing

fixing a cord, he pulls it from time to time, fo as
to imitate the waving motions of the Grous, fitting
on a flexible branch. Experience alfo inftructs
him to turn the head of the *balvanes* againft the
wind when it blows violently; but in ftill weather, he finds it beft to place them oppofite to
each other. If the Grous are driven ftraight
towards his hut, he can difcover by an eafy obfervation, whether they will perch within his
reach. If their flight is irregular, if they fometimes approach and fometimes retire, flapping
their wings, he concludes, that perhaps the whole
flock, or at leaft part of them, will alight near
him. On the other hand, if they fpring not far
from his hut, and fhoot in a rapid fteady courfe,
he is certain that they will pufh on to a diftance.
When the Grous fettle near the hut, the fowler
is informed at leaft thrice by their repeated cries;
he is then cautious not to fire upon them too
fuddenly; he remains ftill in his hut, and without making the leaft noife, allowing the birds
time to examine their fituation, and to quiet their
apprehenfions. When they are fettled and begin
to feed, he takes his aim fteadily, and fires. But
however numerous the flock be, though it even
amount to fifty or a hundred, he can hardly expect to kill more than one or two at each fhot;
for thefe birds do not group together, but commonly perch on a feparate tree, and hence
ftraggling bufhes are better for the fport than a
thick foreft. However, when there is no fnow

lying

lying on the ground, this amufement is fometimes
taken in open ftubble fields, the crops of oats,
barley, buck-wheat, being led, the hut is covered
with ftraw; there the fport is tolerably fuccefsful,
except in fevere weather, when thefe birds are
difperfed and concealed.   But the firft fine day
that fucceeds makes them more eafily caught;
and a fhooter, who has taken his ftation properly,
can, without any affiftance of horfemen, and with
bird-calls alone, entice them to his hut with
eafe.

It is afferted that, when thefe birds fly in
flocks, they are led by an old cock, who con-
ducts them like an experienced chief, and teaches
them to fhun the decoys of the fportfmen; fo
that in this cafe it is exceedingly difficult to
drive them to the *balvane*, and all that can be
then expected is to intercept a few of the
ftragglers.

The proper time for the fport is from fun-rife
to ten o'clock in the forenoon; and from one
o'clock in the afternoon to four.   But in autumn,
when the air is ftill and clofe, it may be continued
without interruption through the whole day; for
the Grous then feldom fhift their place.   And in
this way, they may be chafed from tree to tree,
till near the winter folftice; about that time they
grow more wild, fhy, and cunning; they even
change their accuftomed haunt, unlefs they are
confined by the rigours of the feafon.

It

It is said to be a sign of fair weather, when the
Grous sit on the tops of the trees, and upon the
young shoots; but if they descend to the lower
branches, and squat, it forebodes an approaching
storm. I should not take notice of these remarks
of the sportsmen, if they did not correspond with
the instincts of these birds, which, from what we
have already seen, must be very susceptible of
the impressions made by the varying state of the
atmosphere, and whose sensibility in this respect
may be supposed so great, consistently with probability, as to be affected by the change which
decides the nature of the following day.

When the weather is excessively rainy, they
retire for shelter into the closest and most bushy
forests, and as they are tardy and laborious in
their flight, they can sometimes be hunted down
with dogs, which exhaust them, and catch them
by speed of foot *.

In other countries, the Black Grous is, according to Aldrovandus, caught with a noose; a
net is also used, as has been already observed; but
it would be curious to know the shape, dimensions, and construction of the one with which the
Polish nobleman, of whom Rzaczynski speaks,
caught two hundred and sixty at one time. [A]

* Breslaw's Acts for 1725. This unwieldiness has been remarked by Pliny; and was meant to apply both perhaps to the
Great and the Small Grous.

[A] Specific character of the Black Grous, *Tetrao-Tetrix*: " Its
" tail is forked, its second wing-quills white near the base." Its

egg

egg is yellowifh, fpotted with dark red.   In Lapland, the Black
Grous is taken in fnares ; but formerly it was fhot with arrows.
The people of Siberia have a fingular method for catching thefe
birds during the winter.   They lay a number of poles horizontally
on forked fticks in the open birch forefts, and fet fmall bundles of
corn on them.   At a fhort diftance they plant tall bafkets fhaped
like an invented cone, and place in the mouth of thefe a little
wheel that turns freely on its axis.   The Black Grous are attracted
by the corn, alight on the poles, and after a hafty repaft, fly to the
bafkets, perch upon the rim of the wheel, which, giving way,
precipitates them into the trap.

BROAD-TAILED BLACK GROUS.

Le Petit Tetras a Queue pleine, Buff.

I HAVE, in the preceding article, ſtated the
reaſons which have induced me to refer this
bird to a diſtinct family. Geſner ſpeaks of it
by the name of *Wood-cock, (gallus ſylveſtris)*,
as a bird having red barbils, and a broad undivided tail. He adds, that the male is called
Wood-cock in Scotland, and the female *Grey-hen*.
It is true indeed, that this author, conceiving
that the two ſexes cannot differ much in the colour of their plumage, tranſlates Grey-hen by
gallina fuſca or *Duſky-hen*, in order to bring them
to a nearer conformity; and reſting on this
erroneous verſion, he concludes that this ſpecies
is quite diſtinct from the Mooriſh hen of Turner,
becauſe he imagines this bird is too widely
removed, by the colour of its plumage, from
the male to belong to the ſame family. But
the fact is, that the male is almoſt always entirely black, and the female is nearly the ſame
colour with the gray partridge; and this circumſtance completely decides its identity with the
Black Cock of Scotland; for even Geſner admits,
that in other reſpects they are perfectly alike.
The only difference that I can perceive is, that

o 4 the

the Scotch Black Cock has fmall red fpots under the breaft, the wings, and the thighs; but we have feen in the preceding article, that the young males which in the end become black, are at firft of the colour of the mother, and perhaps the fmall red fpots mentioned by Gefner, are only the traces of their infant plumage before they have acquired the deep jet.

I fee no reafon why Briffon fhould confound this tribe or variety, as he calls it, with the *tetrao,* dotted with white, of Linnæus *; fince one of the characters of this bird, which is termed by the Swedes *rackle-hane,* is its having a forked-tail. Befides, Linnæus gives it no barbils, which, according to the figure and defcription of Gefner, belong to the other birds.

Nor can I fee why Briffon, though he claffes thefe two tribes together, makes only one variety of the forked-tail Black Grous; fince, befides the differences that have been juft noticed, Linnæus exprefsly mentions, that his Grous fprinkled with white, is more fhy and wild, and has a quite different cry; which implies, I fhould imagine, characters deeply impreffed, and more permanent than what conftitute a mere variety.

It would appear therefore more confiftent to diftinguifh thefe into two fpecies of Black Grous; the one including the Scotch Black Cock, and Turner's Moorifh Hen; and the other, characterized by the fmall white fpots under the breaft, and

* *Fauna Suecica,* No. 167.

its

its different cry, would comprehend the Swedifh
rackle-bane. Thus we might reckon four fpecies
of the genus of Grous. 1. The Wood Grous:
2. The Forked-tail Black Grous: 3. The *Rack-
lan*, or *Racklebane*, of Sweden, defcribed by Linnæus: 4. The Moorifh Hen of Turner, or the
Black Cock of Scotland; with flefh barbils on
both fides of the bill, and with an uniform tail.—
Thefe four fpecies are all natives of the northern
climates, and refide either in forefts of pine or of
birch. The third only, or the Swedifh *rackle-
bane*, is the only one that might be confidered as
a variety of the Black Grous, if Linnæus had
not afcertained its having a different note.

M

The BLACK GROUS
WITH VARIABLE PLUMAGE.

Le Petit Tetras à Plumage variable, Buff.

THE Wood Grous are common in Lapland, efpecially when the fcarcity of provifions, or the exceffive multiplication of their numbers, compels them to leave the forefts of Sweden and Scandinavia, and advance into the polar tracts *. Yet they have never been found white in thofe frozen regions; the colour of their plumage feems to be fixed and permanent, and to refift the operation of cold. The fame may be faid of the Little or Black Grous, which are frequent in Courland, and the north of Poland; but Dr. Weigandt, the Jefuit Rzaczynfki, and Klein, affirm that there is in Courland another kind of the fame, termed *White Grous*, which, however, become white only in winter, and by the return of fummer, acquire a reddifh-brown colour, according to Dr. Weigandt; but a bluifh-grey, according to Rzaczynfki. Thefe variations take place generally in both fexes; fo that at all times the individuals have precifely

---

* Klein.

the

the fame colours. They do not perch on trees like the other Grous, but delight in thick brufh-wood and heath ; and generally felect each year a certain fpot, to which they commonly refort when difperfed by fportfmen, by birds of prey, or by the violence of a ftorm. If we hunt them, we ought, when they are firft fprung, to obferve carefully their place of fhelter, fince this will certainly be their rendezvous through-out the year ; and it will be more difficult to fpring them a fecond time, for they will ra-ther fquat on the ground, and endeavour to con-ceal themfelves, in which cafe it will be eafy to fhoot them.

It appears, therefore, that they differ from the Black Grous, not only by their colour, and by the uniformity which obtains between the male and female, but in their habits, fince they never perch. They are alfo diftinguifhed from the ptarmigans, becaufe they inhabit not the lofty mountains, but refide in the woods and among the heaths ; nor are their legs clothed to the toes with feathers. I muft indeed con-fefs, that I would rather have ranked it with the Red Grous, did I not fubmit to the opinion of thefe three intelligent writers, who fpeak of a bird that is a native of their own country.

The HAZEL GROUS.

La Gelinotte, Buff.
Tetrao-Bonafia, Linn. Gmel. &c.
Gallina Corylorum *, Ray, Will. and Klein.

WHAT Varro has faid concerning the Ruftic or Wild Hen, applies fo accurately to the Hazel Grous, that Belon does not hefitate to conclude that they were the fame. It was, according to Varro, a bird uncommonly rare at Rome; and fo difficult to tame, that it could only be raifed in cages, and feldom or never laid eggs in this ftate of captivity. Belon and Schwenckfeld fay the fame of the Hazel Grous; the former conveys, in a few words, a precife notion of the bird, more diftinct than could be given by a long defcription. "Suppofe," fays he, "that you faw a partridge bred by the "croffing of the red with the grey, and having "a few pheafants feathers, and you will have "an idea of the Hazel Grous."

The male is diftinguifhed from the female by a very remarkable black fpot under its throat, and by its orbits, which are of a much deeper red. Their fize is that of the *bartavelle;* the

* In German, *Hafel-buhn*; in Swedifh, *Harpen*; in Polifh, *Jarzabek.*

extent

THE SPOTTED GROUS.

extent of their wings is only twenty-one inches, and hence they fly flowly and laborioufly, and a great effort is required to raife them from the ground: however, they run exceedingly faft *. They have twenty-four quills, that are almoft all equal, in each wing, and fixteen in the tail. Schwenckfeld fays, that there are only fifteen; but he is miftaken, and the lefs excufable, as there is perhaps not a fingle bird that has an odd number of tail quills. The tail is marked near its extremity by a broad blackifh bar, interrupted only by the two middle quills. I fhould not take notice of that circumftance, were it not to confirm the remark of Willughby, that in moft birds the two middle quills do not follow the diftance of the lateral ones, but fometimes project beyond them, and fometimes extend not fo far; fo that in this cafe, the interruption of colour appears to depend on the difference of their pofition.—Like the other Grous, their orbits are red; the toes indented on the fides, though more flightly; the nail of the middle toe, fharp and flat; the legs clothed with feathers before, but only as far as the middle of the tarfus; the gizzard mufcular; the alimentary canal thirty inches and odd; the appendices or *cæca* thirteen or fourteen inches, ftriped with furrows †; the flefh white when dreffed, but more fo within than

* Gefner.   † Willughby.

without,

without, and thofe who have examined it attentively pretend that they can diftinguifh four different colours; in the fame manner as three
different flavours are found in the buftards and
common grous. However, the flefh of the
Hazel Grous is excellent; and hence is derived,
it is faid, its Latin name *Bonafa*; and alfo the
Hungarian appellation *Tfchafarmadar*, or *Cæfar's
Bird*, fignifying that it was fit to be kept for the
Emperor. It is indeed highly efteemed, and
Gefner remarks that it is the only difh fuffered
to appear twice at princes' tables.

In the kingdom of Bohemia, it is as much
eaten at Eafter, as lamb in France; and it is
cuftomary to fend it in prefents from one perfon
to another *.

The Hazel Grous lives, both in fummer and
winter, on nearly the fame food as that of the
Common Grous. We find in their ftomach, in
the fummer, the berries of the fervice-tree, of
the bilberry, the bramble, and the heath; the
feeds of the Alpine elder, the pods of the *fal-
tarella*, the catkins of the birch and of the
hazel, &c.; and in winter, we meet with
juniper berries, the buds of the birch, the tops
of heath, fir, juniper, and of fome other ever-
greens †. When the Hazel Grous is kept in
confinement, it may be fed with wheat, barley,

* Schwenckfeld.
† Ray, Schwenckfeld, and Rzaczynfki.

and

and other grain; but, like the Common Grous, it does not long furvive the lofs of its liberty *; whether becaufe it is fhut up fo clofely as to affect its health, or that its favage, or rather generous nature, will not brook the flighteft encroachment on its freedom.

The time of fport returns twice a-year, in fpring and in autumn; but the latter feafon is the moft favourable. They are attracted by the found of bird-calls which imitate their note, and horfes are led into the field, becaufe it is a vulgar opinion, that the Hazel Grous are fond of thefe animals †. It has alfo been remarked by the fportfmen, that if the cock be firft caught, the hen feeks her mate with anxious folicitude, and returns feveral times to vifit the fpot, with other males in her train; but if the hen be firft enfnared, the cock joins another family, and totally forgets his former attachments ‡. Certain it is, that when one of thefe birds is furprifed and roufed, it fprings, making a loud noife, and, perching on a tree, it fits motionlefs and unconcerned, while the fportfman meditates its deftruction. Commonly they fettle on the centre of the tree, where the boughs part from the trunk.

As much has been faid of the Hazel Grous, many fables have been told: the moft abfurd

---

* Gefner and Schwenckfeld.
‡ Gefner.   ‡ Ibid.

are

are thofe concerning its manner of propagating.
Encelius, and others, affert that they copulate
with their bills, that the cocks themfelves lay
when they grow old, and that their eggs being
hatched by the toads, produce wild bafilifks, in
the fame manner as the eggs of the common
cocks, if hatched by toads, give birth to the domeftic bafilifks. And left we fhould entertain
fufpicions with regard to thefe bafilifks, Encelius defcribes one that he faw *; but unfortunately he neither tells us whether he beheld it emerge from the egg, or beheld this
egg excluded by the male. Moft of thefe abfurdities take their rife from the mifreprefenta-
tion of facts; and it is probable that the Hazel
Grous bill like the turtle doves, and toy with
each other to raife the fwell of love.

According to the opinion of fportfmen, the
Hazel Grous comes in feafon in the months of
October and November; and at that time the
males only are killed, being decoyed by a
kind of whiftling analogous to the fhrill note of
the females; they haften to the fpot, making a
loud ruftling noife with their wings, and are
fhot as foon as they alight.

The females, like other large birds, form
their neft on the ground, and commonly conceal it under hazels, or below the fhade of a

* In Gefner.

broad

broad mountain fern. They commonly lay
twelve or fifteen eggs, and fometimes even
twenty, and thefe are fomewhat larger than
pigeons eggs *. They fit three weeks, and
have feldom more than feven or eight young †,
which run as foon as they are hatched, as ufual
in moft of the fhort-winged birds. As foon as
the young are able to fly, the parents remove
from the tract where they bred; and being thus
forfaken, they pair and difperfe, to form new
fettlements, and in their turn to fend off other
colonies ‡.

The Hazel Grous delight in forefts, where
they can find their proper fuftenance, and conceal themfelves from the rapacious birds, which
they dread exceedingly, and perch, for fhelter,
on the low branches ||. Some affirm that they
prefer the mountain forefts; but they alfo
inhabit the woods that grow in the plains,
for they are plentiful in the neighbourhood
of Nuremberg. They are frequent alfo in
the woods that clothe the bottom of the Alps
and the Apennines. They are found likewife in the mountain of Giants in Silefia,
in Poland, &c. Anciently they were fo
numerous, according to Varro, in a little
ifland in the Liguftic Sea, now the Gulph

* Schwenckfeld. † Frifch.
‡ Gefner. || Ibid.

of Genoa, that it was called the *Ifland of the Hazel Grous*. [A]

[A] Specific character of the Hazel Grous, *Tetrao-Bonafia.* " Its tail-quills are cinereous, with black dots, and a black " ftripe; except the two intermediate." It is larger than the Englifh partridge. It occurs in many parts of the north of Europe; in Ruffia, Siberia, and Lapland. It has a fhrill piping note, and may be decoyed by imitating the found.

M

The SCOTCH HAZEL GROUS.

IF this bird be the fame with the *Gallus
palustris* of Gefner, as Briffon thinks, the
figure which the German naturalist gives, muft
undoubtedly be very inaccurate, fince no feathers
are reprefented on the legs; and, on the other
hand, red barbils appear under the bill. Is it
not natural then to fufpect that this figure belongs to a different bird? However, the *Wood-
cock*, or *Cock of the Marfh*, is excellent meat;
and all that we know of its hiftory is, that it
delights in wet fituations, as its name denotes.
The Authors of the *British Zoology* fuppofe, that
what Briffon takes for the *Scotch Hazel Grous*,
is really the *Ptarmigan* in its fummer garb, and
that its plumage becomes almoft always white
in winter. But if this were the cafe, it muft
alfo lofe the feathers which cover the toes; for
Briffon exprefsly notices, that it is only clothed
to the origin of the toes, and the ptarmigan in
the British Zoology is feathered even to the
nails; befides, thefe two birds, as they are reprefented in the Zoology, and in Briffon's
work, refemble each other neither in appearance nor ftructure. Briffon's Scotch Hazel
Grous is fomewhat larger than ours, and its

tail

tail fhorter; it refembles that of the Pyrenees
in the length of its wings; its legs clothed
before with feathers as far as the origin of the
toes; in the length of the middle toe compared
with thofe on the fides; and in the fhortnefs of
the hind toe: it differs, becaufe its toes are not
indented, and its tail has not the two long narrow feathers, which is the mott obvious character of the Pyrenean Hazel Grous. I need
take no notice of the colours of the plumage,
the figure will convey a clearer idea than any
defcription; and befides, nothing is more uncertain, fince they vary confiderably in the fame
individual at different feafons.

M

THE PIN-TAILED GROUS.

The PIN-TAILED GROUS.

La Ganga, vulgairement La Gelinotte des Pyrenees *, Buff.
Tetrao-Alchata, Linn. Gmel. Klein, &c.
Bonafa Pyrenaica, Briff.
The Partridge of Damafcus, Will. and Ray.
The Kitiwiah, or African Lagopus, Shaw.

THOUGH there is a wide difference between
words and things, it often happens in natural hiftory, that the mifapplication of terms is
the fource of multiplied miftakes; we have
therefore made it an invariable rule, to difcover, as much as poffible, the true meaning of
names.

Briffon, confidering the Damafcus or Syrian
Partridge of Belon, as the fame fpecies with his
Pyrenean Hazel Grous, ranges it among the
appellations beftowed in different languages on
that tribe, and quotes Belon as his authority for
the Greek name Συροπερδριξ. But he is miftaken in two points: Firft, Belon tells us himfelf,
that the bird which he calls *the Damafcus Partridge*, is a different fpecies from what authors term
Syroperdrix, which has a black plumage and a

---

* *i. e.* The Ganga, commonly called the Pyrenean Ptarmigan. In Turkifh, *Kata*; in Spanifh, *Ganga*.

red

red bill. Secondly, Briſſon, writing the word in Greek characters, ſeems to inſinuate that it is derived from that language, while Belon poſitively mentions that it is originally Latin. Laſtly, it is difficult to conceive what led Briſſon to conſider the *ænas* of Ariſtotle as the ſame ſpecies with his Pyrenean Hazel Grous; for Ariſtotle claſſes his *ænas*, which is the *vinago* of Gaza, with the pigeons, the turtles, and the ring-doves, (in which he is followed by all the Arabians,) and he expreſsly mentions that, like theſe birds, it only lays two eggs at a time. But we have already ſeen that the Hazel Grous lays a much greater number; and conſequently the *ænas* of Ariſtotle cannot be conſidered as the Pyrenean Hazel Grous, and ought therefore to be referred to a different ſpecies.

Rondelet conceived, that the Greek word was not οιναϛ, but ought to be read ιναϛ, whoſe primitive ſignifies a *fibre* or *thread* *; becauſe the fleſh is ſo fibrous and hard that it muſt be flead before it can be eaten. But if it were really the ſame bird with the Pyrenean Hazel Grous, we might adopt the correction of Rondelet, and yet give to the word *inas* a more happy explanation, and more conſiſtent with the genius of the Greek language, which paints whatever it would expreſs; if we conceive it to denote the two threads or narrow feathers of the

* ις, ινος.

tail

tail of the Pyrenean Hazel Grous, and which is their characteriftical diftinction. But unfortunately Ariftotle does not fay a word concerning thefe threads, which had efcaped his obfervation; nor does Belon take any notice of this circumftance in his defcription of the Damafcus partridge. Befides, the name οινας*, or *vinago*, is more fuitable to this bird, as it arrives in Greece about the beginning of autumn, which is the feafon of vintage; for the fame reafon that in Burgundy a certain kind of thruſhes are called by the people in that county *vinettes*.

It follows from what has been faid, that the *fyroperdrix* of Belon, and the *ænas* of Ariftotle, are not the Pyrenean Hazel or Pin-tailed Grous, any more than the *alchata, alfuachat,* and the *filacotona*, which appear to be Arabian names, and certainly denote a bird of the pigeon kind.

On the other hand, the Syrian bird, which Edwards terms *the little heath-cock, with two thread-like feathers in the tail,* and which the Turks call *kata,* is really the fame with the Pyrenean Hazel-Grous. This author tells us, that Dr. Shaw names it *kittawiah,* and that he only gives three toes to each foot; but he alleges that the traveller has committed this overfight in not attending to the hind toe, which is hid under the plumage of the legs. Yet he had a little before mentioned, (and we readily

---

* From οινος, wine.

perceive

perceive it from the figure,) that the fore-part
only of the leg is covered with white feathers
like hairs; and it is difficult to conceive how
the hind toe could be concealed under the anterior plumage. It would be more natural to
fay, that it efcaped Dr. Shaw's obfervation, by
its diminutive fize, for it is only two lines long.
The two lateral toes are alfo very fhort compared with the middle one, and in them all, the
edges are marked with fmall indentings, as in
the common Grous. The Pin-tailed or Pyrenean Hazel Grous, feems therefore to be quite
a diftinct fpecies from the true Hazel Grous.
For, 1. its wings are much broader in proportion to the reft of its body, and confequently it
muft fly fmoothly and rapidly, and have habits
different from thofe of tardy birds. 2. We learn
from the obfervations of Dr. Ruffel, quoted by
Edwards, that it flies in numerous flocks, and
fpends the greateft part of the year in the deferts
of Syria, and does not venture near the city of
Aleppo, except in the months of May and June,
when it is obliged to refort to places where it can
get water. We know too that the Hazel Grous
is a timorous bird, and never deems itfelf fecure
from the vultures talons, unlefs concealed in the
moft fhady trees. The Pin-tailed Grous, which
the inhabitants of Catalonia call the *partridge of
Garrira* *, is nearly the bulk of the gray partridge;

* Barrere, *Ornithologia.*

13   the

the orbits are black, nor are the eye-brows red or flame coloured; the bill is almoft ftraight; the noftrils are placed at the bafe of the upper mandible, and joining the feathers which cover the face; the fore-part of the leg is feathered to the origin of the toes; the wings are of confiderable length, and the fhafts of the quills are black; the two quills in the middle of the tail are twice as long as the reft, and very narrow where they project; the lateral quills grow fhorter and fhorter until the laft one. We may remark that of all thefe properties which characterife the pretended Hazel Grous of the Pyrenees, there is not one which exactly agrees with the Hazel Grous [*].

The female is of the fame fize with the male, but differs by its plumage, the colours of which are fainter, and by the filaments in the tail, which are not fo long. It appears that the male has a black fpot under its throat, and that the female, inftead of this, has three rings of the fame colour, which encircle its neck like a collar.

I fhall not attempt to defcribe the colours of the plumage; I fhall only obferve that they have a great affinity to thofe of the bird known at Montpelier by the name of *angel*, of which John Culmann communicated a defcription to Gefner [†]; but the two long feathers of the tail

feem

[*] Edwards and Briffon.

[†] " The feathers are of a dufky colour inclining to black and " yellowifh, verging on rufous," fays Gefner, fpeaking of the *angel*.

" Variegated

feem to be omitted in this defcription, and alfo in
the figure fent by Rondelet to Gefner, of this
fame bird, which he had taken for the *œnas* of
Ariftotle.   In fhort, there feems to be reafon to
doubt the identity of thefe two fpecies, notwithftanding their correfpondence in the plumage and
in the place of refidence ; unlefs we fuppofe that
the fubjects defcribed by Culmann and defigned
by Rondelet were females, in which the threads
of the tail were much fhorter, and confequently
lefs remarkable.

   This fpecies is found in moft of the warm
countries in the ancient continent ; in Spain,
in the fouth of France, in Italy, in Syria, in
Turkey and Arabia, in Barbary, and even at
Senegal ; for the bird figured in the *Planches
Enluminées* by the name of the Senegal Hazel
Grous, is only a variety, and fomewhat fmaller,
but has the fame long feathers or threads in the
tail the lateral quills become gradually fhorter
the farther they are placed from the middle, the
wings are very long, the legs covered before
with a white down, the mid-toe much longer
than thofe near the fides, and the hind one exceedingly fhort; laftly, it has no red fkin over
the eyes, and differs from the Pin-tailed Grous
only in being rather fmaller, and its plumage
deeper tinged with reddifh.   It is therefore only

   " Variegated with olive, yellowifh black, and rufous," fays
Briffon, in his defcription of the Pyrenean Hazel Grous.

                                          a variety

a variety of the fame fpecies, produced by the influence of climate ; and what ought to fhew that this bird is different from the Hazel Grous, and fhould therefore be diftinguifhed by a different name, is, that befides the difparity of figure, it always inhabits the warm countries, and never occurs in the cold or even the temperate climates ; whereas the Hazel Grous are rare except in chilly tracts.

It may be proper in this place to tranfcribe what Dr. Shaw informs us with refpect to the *Kittawiah*, or Barbary Hazel Grous, and which is all we know on the fubject, that the reader may compare it with the Pin-tailed Grous, or the Pyrenean Hazel Grous, and judge if they are really two individuals of the fame fpecies.

" The *Kittawiah* or *African Lagopus* *, (as we
" may call it,) is another bird of the gregarious
" and granivorous kind, which likewife wanteth
" the hinder toe. It frequenteth the moft barren,
" as the *Rhaad* doth the moft fertile parts of thefe
" countries, being in fize and habit of body like
" the dove, fhort feathered feet alfo, as in fome
" birds of that kind. The body is of a livid co-
" lour, fpotted with black ; the belly black-
" ifh ; and upon the throat there is the figure of
" a half moon, in a beautiful yellow. The tip of
" each feather of the tail hath a white fpot upon

* This name is improperly applied, fince the bird is not feathered under the toes.

" it,

" it, and the middle is long and pointed, as in the
" *Merops.* The flesh is of the same colour with
" the *Rhaad's*, red upon the breast, and white in
" the legs, agreeing further in being not only
" of an agreeable taste, but easy digestion."
Shaw's Travels, p. 253. [A]

[A] Specific character of the Pin-tailed Grous, *Tetrao-Alchata.* " Above variegated, the two middle tail-quills twice as long and subulated."

M

The RED GROUS*.

L'*Attagas*, Buff.
Tetrao-Lagopus, var. 3d. Gmel.
Bonasa Scotica, Briss.
Tetrao Scoticus, Lath.
Attagen, Frif.
The *Moor-Cock*, or *Moor-Fowl*, Sibb.
The *Red-Game*, *Gorcock*, or *Moor-Cock*, Will.

THIS is Belon's *francolin*, which we muft not confound, as fome ornithologifts have done, with the *francolin* defcribed by Olina. Thefe are two birds widely different both in their form and in their habits: the laft delights in plains and low fituations; it has not the beautiful flame-coloured orbits, that give the other fo diftinguifhed an appearance; its neck is fhorter and its body thicker; the feet reddifh, furnifhed with fpurs, and not feathered, as its toes are not indented; in fhort, it bears no refemblance at all to the bird which we at prefent confider.

The ancients have faid a great deal about the *attagas*, or *attagen* (for they ufe both names indifferently). Alexander the Myndian tells us in his commentary on Athenæus, that it was rather larger than a partridge, and its plumage, which was of a reddifh ground, was mottled with fe-

* In Greek, Ατλαγην or Ατλαγας.

veral

veral colours. Ariftophanes had faid nearly the
fame thing; but Ariftotle, according to his
commendable cuftom of marking the analogy
between unknown objeĉts and fuch as are common, compares its plumage to that of the wood-
cock (σκολοπαξ). Alexander the Myndian fubjoins,
that its wings are fhort, and its flight tardy; and
Theophraftus remarks that, like the other heavy
birds, as the partridge, the cock, the pneafant,
&c. it is hatched without feathers, and can run
as foon as it quits the fhell. Like thefe alfo, it
welters in the duft *, and feeds on fruits, devour-
ing the berries and grain which it finds, fometimes eating the plants themfelves, fometimes
fcraping the earth with its nails; and as it runs
more than it flies, it was cuftomary to hunt it
with dogs, and this chafe was fuccefsful †.

Pliny, Ælian, and others fay, that thefe birds
lofe their cry with their liberty; and that, owing
to the depreffion of their native faculties, they
are very difficult to tame. Varro, however, in-
ftruĉts us how to breed them, and the mode is
nearly the fame with that of raifing peacocks,
pheafants, Guinea fowls, partridges, &c.

Pliny informs us, that this bird, which had
been very rare, was become more common in

* The ancients called thefe birds *Pulveratrices,* which roll in
duft to rid themfelves of the infeĉts that torment them; in the
fame manner as the aquatic fowls feek to remove them by fprink-
ling water on their wings.

† Oppian *in Ixeuticis.* This author adds, that they are fond
of ftags, and, on the contrary, have an averfion to cocks.

8   his

his time; that it was found in Spain, in Gaul,
and on the Alps; but that thofe from Ionia were
the moft efteemed. In another place, he tells
us, that there were none in the ifland of Crete.
Ariftophanes fpeaks of thofe found in the vicinity
of Megara in Achaia. Clement of Alexandria
fays, that thofe from Egypt were reckoned the
moft delicious by the epicures. Some there
were alfo in Phrygia, according to Aulus Gellius,
who defcribes it as an Afiatic bird. Apicius directs us how to cook the Red Grous, which he
joins with the partridge; and St. Jerome mentions it in his letters as a moft exquifite difh *.

However, to judge whether the *attagen* of
the ancients is our Red Grous, we muft collect
its hiftory from the writings of the moderns, and
form the comparifon.

It appears that the word *attagen*, though with
various corruptions, has generally been ufed by
the modern authors who have written in Latin,
as the name of this bird †. It is true indeed, that
fome ornithologifts, as Sibbald, Ray, Willughby,
and Klein, have referred it to the *lagopus altera* of
Pliny ‡; but, befides that Pliny only mentions
it by the way, and fo curforily as to give no pre-

* " You fmell of Grous *(Attagen)*, and yet boaft of eating
" Goofe," faid St. Jerome to an hypocrite, who pretended to live
on fimple diet, but in private regaled himfelf with delicacies.

† *Attago, Attago, Atago, Atchemigi, Atacuigi, Tagenarios, Tagi-
aari,* are all words corrupted from *Attagen.* Gefner.

‡ Hift. Nat. lib. x. 48.

cife

cife idea, is it likely that this great naturalift, who had treated at great length on the *attagen* in the fame chapter, would fay a few words of it afterwards under another name, and without giving notice ? This reflection is alone fufficient, in my opinion, to prove that the *attagen* of Pliny and his *lagopus altera* were different birds; and we fhall afterwards know what they really were.

Gefner was told, that this bird is commonly called *franguello* at Bologna; but Aldrovandus, who was a native of that place, tells us, that the name *franguello (binguello*, according to Olina), was given commonly to the chaffinch, and which is evidently derived from the Latin *fringilla*. Olina fubjoins, that in Italy, his *francolin*, which we have faid is a bird different from ours, was generally named *franguellina*; a word corrupted from *frangolino*, and to which a feminine termination was added, to diftinguifh it from *fringuello*.

I know not why Albin, who has copied the defcription that Willughby gives of the *lagopus altera* of Pliny, has changed the name into *Cock of the Marfh*; unlefs becaufe Tournefort fays, that the Samian *francolin* haunts marfhes. But if we compare the defcriptions with the figures, we fhall readily perceive that the Samian *francolin* is entirely different from the bird which Albin, or his tranflator, has been pleafed to term

Cock

If we compare the accounts of the moderns
in regard to the Red Grous, with what the ancients have faid on the fame fubject, we find
the former more accurate and full; yet we have
ftill facts enow from which we may conclude
the identity of that bird with the *attagen* of antiquity.

To conclude, though I have been at pains to
remove the confufion in which this fubject is
involved, and to affign to each fpecies the
characters that have been indifcriminately beftowed, I cannot expect that I have been
equally fuccefsful in clearing every point. The
uncertainty which clouds our views, is owing
entirely to that latitude in the ufe of names in
which naturalifts have indulged themfelves, and
which throws obftacles almoft infurmountable
on every attempt to connect our prefent information with the difcoveries of paft ages. [A]

[A] Specific character of the Red Grous, *Tetrao Scoticus*,
LATH.—" It is ftriated tranfverfely with rufous and blackifh; its
" fix exterior tail-quills on either fide, blackifh." Mr. Pennant
thinks that this bird is peculiar to Britain. It occurs in Wales, and
in the north of England, and is numerous in the Highlands of
Scotland. Its egg is elongated; tawny, marked with irregular
blood coloured blotches, having dots interfperfed.

The WHITE ATTAGAS.

IT is found in the mountains of Switzerland, and in thofe around Vicenza. I have nothing to add to what has been faid in the preceding article, except that Gefner's fecond fpecies of *lagopus* appears to be really one of thefe birds, though the white of its plumage is pure only on the belly and the wings, and is clouded with brown or black on the reft of the body: for we have already feen that the colour of the male is not fo deep as that of the female; and we know that in moft young birds, and particularly of this kind, it never acquires its due intenfity till the fecond year. Alfo Gefner's defcription fuits this fpecies exactly; the eyebrows red, naked, curved, and prominent; the feet feathered as far as the nails, but not below; the bill fhort and black; the tail alfo fhort; its refiding in the Swifs mountains, &c. I fhould imagine that this bird was really a White Attagas, a male, and young, weighing only fourteen ounces inftead of nineteen, the ufual bulk.

I would draw a fimilar conclufion with regard to Gefner's third fpecies of *lagopus*, which feems to be the fame with what the Jefuit

Rzaczynfki

Rzaczynſki mentions by the Poliſh name *parowa*. In both, a part of the wings and belly is white, the back and the reſt of the body of a variegated colour; their feet feathered; their flight laborious; their fleſh excellent; and their ſize equal to that of an ordinary hen. Rzaczynſki takes notice of two kinds; the one ſmall, which I am at preſent conſidering; the other larger, and which is probably a ſpecies of the Hazel Grous. This author ſubjoins, that both birds are found entirely white in the Palatinate of Novogorod. I do not claſs them with the Ptarmigan, as Briſſon has done Geſner's ſecond and third ſpecies of *lagopus*; becauſe their feet are not feathered beneath, which is the moſt ancient and decided character.

The PTARMIGAN*.

Le Lagopede, Buff.
Tetrao-Lagopus, Linn. Gmel. &c.
Lagopus, Pliny.
Tetrao Mutus, Martin.
White Game, Will.

THIS bird has been called the *White Par-tridge,* very improperly; since it is not a partridge, and is white only in winter, on account of the intense cold to which it is exposed during that season on the lofty mountains of the North, which it commonly inhabits. Aristotle, who was unacquainted with the Ptarmigan, knew that partridges, quails, swallows, sparrows, ravens, and even hares, stags, and bears, suffer, in similar situations, the same change of colour †. Scaliger adds the eagles, vultures, sparrow-hawks, kites, turtle doves, and foxes ‡; and it would be easy to increase this list, by the names of many birds and quadrupeds on which cold can produce similar effects. We may therefore infer, that the white colour is not

---

* In Norwegian, *Rype*; in Iceland, the cock is called *Riup-karre,* and the hen, *Riupa.*
† *De coloribus,* cap. vi. and *Hist. Anim.* lib. iii. 12.
‡ *Exercitationes in Cardanum.*

permanent,

THE PTARMIGAN.

permanent, and cannot be confidered as a difcriminating character of the Ptarmigan; efpecially as many fpecies of the fame genus, as the
Little White Grous, according to Rzaczynfki *
and Dr. Weygandt †, and the White Attagas,
according to Belon, are liable to the fame variations of colour. It is aftonifhing that Frifch
was not informed, that his *White Mountain
Francolin*, which is the Ptarmigan, is fubject to
this influence of cold; for if he was acquainted
with this fact, it is equally ftrange that he has
omitted to mention it. He only fays, that he
was told that no White Francolins could be met
with in fummer; and therefore he tells us, that
they were fometimes found (in fummer no
doubt) with their wings and back brown, but
which he had never feen. This was the place,
therefore, where he ought to have added, that
they are white only in winter, &c.

Ariftotle, as I have already faid, was unacquainted with the Ptarmigan; what demon-
ftrates the affertion, is a paffage in his Natural
Hiftory, where he fays, that the hare is the only
animal whofe feet are covered with fur on the
fole; but, if he had known the fact, he would
certainly not have omitted, in a place where he
draws general comparifons, to mention a bird
that is diftinguifhed by the fame property.

* *Auctuarium Poloniæ.*   † Breflaw's Acts, Nov. 1725.

The

The name *Lagopus* is that which Pliny and other writers of antiquity have beſtowed on this ſpecies of birds. The moderns have there-fore committed an impropriety, when they have applied a word which marks the diſtinguiſhing character of the Ptarmigan to the nocturnal birds, whoſe feet are feathered above and not below *. Pliny adds, that it is as large as a pigeon, that it is white, that it is excellent, and that it reſides on the ſummits of the Alps; laſtly, that it is ſo wild that it can hardly be reduced to the domeſtic ſtate; and he concludes with telling us, that its fleſh ſoon runs into putre-faction.

The laborious accuracy of the moderns has completed this ſketch of antiquity. They have noticed that glandulous ſkin which forms a ſort of red eye-brows, but of a brighter colour in the male than in the female; it is alſo ſmaller in the latter, and the two black ſtreaks are want-ing on the head, which in the male ſtretch from the bottom of the bill to the eyes, and even extend near the ears. Except in this cir-cumſtance, the male and female are perfectly alike in their external form; and all that I ſhall afterwards mention on this ſubject will apply to both equally.

The ſnowy colour of the Ptarmigan is not ſpread over its whole body, but is ſtained even

* Belon, Willughby, and Klein.

in

in winter. This exception obtains efpecially in the quills of the tail, which are black, with a little white at the point; nor does it appear from the defcription, that this colour tinges continually the fame quills. Linnæus, in his *Fauna Suecica*, defcribes the middle ones as black; and in his *Syftema Naturæ*, he fays, with Briffon and Willughby, that thefe are white, and the lateral quills black. Thefe naturalifts feem not to have examined their fpecimens with fufficient accuracy. In the individual which I have caufed to be figured, and in others which I have viewed, I found the tail compofed of two rows of feathers, one over the other, the upper one entirely white, and the under one black, and each confifting of fourteen feathers *. Klein takes notice of a bird, which he received from Pruffia on the 20th of January 1747, and which was perfectly white, except the bill, the lower part of the tail, and the fhafts of fix quills of the wings. The Lapland prieft, Samuel Rheen, whom he quotes, fays, that the Snow Fowl, or Ptarmigan, has not a fingle black feather, except the female, which has one of that colour in each wing. And the white partridge, of which Gefner fpeaks, was indeed entirely white, except round the ears, where

* Thefe cannot be counted exactly without plucking, as we have done, above and below the rump; it was in this way that we afcertained that there were fourteen white above, and as many black below.

there

there were fome black marks; the coverts of
the tail, which are white, and extending its whole
length, conceal the black feathers, are what
have occafioned moft of thefe miftakes. Briffon
reckons eighteen quills in the tail, while Willughby, and fome other ornithologifts, reckon
only fixteen; and there are really only fourteen. It would feem, that the plumage of this
bird, how variable foever, is more uniform than
the naturalifts reprefent it *. There are twenty-
four quills in the wings, the third one, reckoning from the outer fide, is the longeft, and the
firft fix have black fhafts, though the webs are
white: the down which fhade the feet and toes
as far as the nails, is very thick and foft; and it
has been faid, that this is a kind of fur-gloves
which nature has given to thefe birds, to defend
them from the intenfe cold of their native climates. The nails are very long, even that of

---

* It is not furprifing, that authors differ about the white or
black colour of the lateral tail-feathers of this bird; for in fpreading out the tail with the hand, it is eafy to terminate the fides
either by the black or the white feathers. Daubenton the
younger has well remarked, that there is another method of
fettling the contradiction of authors, and of fhewing clearly
that the tail confifts only of fourteen quills all black, except
the outer one, which is edged with white near its origin, and
the tip, which is white in them all; becaufe the fhafts of thefe
fourteen black quills are twice as thick as the fhatts of the fourteen white quills, and do not project fo far, not over-lapping
entirely the fhafts of the black quills; fo that we may regard
thefe white feathers as only coverts, though the four middle ones
are as large as the black ones, which are all very nearly equal in
length.

the

the little hind toe; that of the mid-toe is
fcooped lengthwife, and its edges are fharp,
which enables it to form holes in the fnow with
eafe.

The Ptarmigan is at leaft as large as a tame
pigeon, according to Willughby; its length is
fourteen or fifteen inches, the extent of its
wings twenty two inches, and its weight fourteen ounces: ours is rather fmaller. But Linnæus remarks, that they are of different fizes,
and that the fmalleft inhabits the Alps *. He
fubjoins, indeed, that the fame bird is found in
the forefts of the northern countries, and efpecially in Lapland; which gives room to fufpect,
that this fpecies is different from our Alpine
Ptarmigan, which has different habits, and prefers the lofty mountains: unlefs perhaps we fay,
that the cold which prevails on the fummits of
the Alps is nearly the fame with what is experienced in the vallies and forefts of Lapland. But
the difagreement of writers with refpect to the
cry of the Ptarmigan feems to prove decidedly,
that there is a confufion of fpecies. Belon
fays, that it has the note of the partridge: Gefner, that the voice fomewhat refembles that
of a ftag; Linnæus compares it to a prattling
and jeering. Laftly, Willughby fpeaks of the
feathers on its feet as a foft down *(plumulis
mollibus)*; and Frifch compares them to hogs

* Fauna Suecica.

briftles.

briftles. But how can we reconcile fuch oppofite qualities, how refer fuch different characters to the fame fpecies? There is reafon
then for the divifion which I have drawn between the Ptarmigans of the Alps, the Pyrennees, and fuch other mountains, and the birds
of the fame genus that occur in the forefts, and
even in the plains of the northern regions.

We have already feen, that in winter the
Ptarmigan is robed in white; in fummer, it
is covered with brown fpots, which are fcattered irregularly on a white ground. It may
be faid, however, never to enjoy the folftitial
warmth, and to be determined by its fingular
ftructure to prefer the chilling froft; for as the
fnow melts on the fides of the mountains, the
bird conftantly afcends, till it gains the fummits,
where reigns eternal winter. It would feem to
be oppreffed by the dazzle of the folar rays;
it withdraws from the luftre of day, and forms
holes and burrows under the fnow. It were
curious to inveftigate the internal and intimate
ftructure of the Ptarmigan, and difcover the
reafon why cold feems fo neceffary to its exiftence, and why it fo carefully fhuns the prefence
of the fun; while almoft every animated being
longs for his return, and hails his approach,
as the father of Nature, the fource of delight,
whofe benign influence infpires and enlivens
all. Muft we afcribe it to the fame caufes
                                    which

which make the nocturnal birds retire from his effulgence? or is the Ptarmigan the *kakkerlac* of the winged tribe?

Such a difpofition, however, will evidently render this bird difficult to tame, and Pliny exprefsly mentions the fact *. Yet Redi fpeaks of two Ptarmigans, which he calls *White Partridges of the Pyrenees*, that were bred in the volery of the garden at Boboli, belonging to the Grand Duke.

The Ptarmigans fly in flocks, but never foar aloft; for they are heavy birds. When they perceive any perfon, they remain ftill on the fnow to avoid being feen; but they are often betrayed by their whitenefs, which furpaffes the fnow itfelf. However, whether through ftupidity or inexperience, they are foon reconciled to the fight of man; they may often be caught by prefenting bread, or a hat may be thrown before them, and a noofe flipped round the neck, while they are engaged in admiring this new object; or they may be difpatched by the blow of a ftick behind †. It is even faid, that they will not venture to pafs a row of ftones rudely piled like the foundation of a wall, but will conftantly travel clofe by the fide of this humble barrier, quite to the fpot where the fnares are placed.

* Coll. Acad. Part Etrang. i. 520.   † Gefner.

They

They live upon the buds and tender ſhoots of
the pine, the birch, the heath, whortle berry, and
other Alpine plants *. It is to the nature of their
food undoubtedly, that we muſt aſcribe the ſlight
bitterneſs of their fleſh †, which otherwiſe is excellent for the table; it is dark-coloured, and is
a very common ſort of game in Mount Cenis,
and in all the towns and villages near the mountains of Savoy ‡.   I have eaten of it, and found
it had much the flavour of hare.

The females lay and hatch their eggs on the
ground, or rather on the rocks §;—this is all that
we know with regard to their propagation. We
ſhould require wings to ſtudy the inſtincts and
habits of birds, eſpecially of thoſe that will not
bend to the yoke of domeſtication, and which
delight in deſerts.

The Ptarmigan has a very thick craw, and a
muſcular gizzard, in which ſmall ſtones are
found mixed with its aliments.   The in-
teſtines are thirty-ſix or thirty-ſeven inches
long; the *cæca* are thick, fluted, and very
long, but not uniform, and are, according to
Redi, full of minute worms‖; the coats of the
ſmall inteſtine are covered with a curious net-
work, formed by a multitude of ſmall veſſels, or
rather of little wrinkles diſpoſed regularly ¶.   It
has been obſerved that its heart is ſomewhat

* Willughby & Klein.   † Geſner.   ‡ Belon.
§ Geſner & Rzaczynſki.   ‖ Coll. Acad. Part. Etrang. tome i.
¶ Klein & Willughby.

ſmaller

fmaller, and its fpleen much fmaller, than in the
Red Grous *; and that the cyftic and hepatic
ducts join the inteftines, at a confiderable diftance
from each other †.

I cannot clofe this article without obferving
with Aldrovandus, that Gefner joins to the different names which have been given to the
Ptarmigan, that of *urblan*, conceiving it to be an
Italian word ufed in Lombardy; yet this term
is totally unknown, both in the language of Italy,
and in that which is fpoken in Lombardy. The
fame perhaps may be faid of the words *rhoncas*
and *herbey*, which, according to the fame author,
the Grifons, who fpeak Italian, beftow on the
Ptarmigan. In the part of Savoy which borders
on the Valais, it is called *arbenne*, which, being
corrupted by the pronunciation of the Swifs and
Grifon peafants, might pafs changed into fome
of the words juft mentioned. [A]

* Roberg *apud Kleinum*.
† Redi, Collect. Acad. Part. Etrang. tome i.

[A] Specific character of the Ptarmigan, *Tetrao-Lagopus :* " It
" is cinereous ; its toes fhaggy, its wing-quills white ; its tail-quills
" black and white at the tip, the intermediate ones white." The
Ptarmigan occurs fometimes in the colder parts of England, and is
pretty frequent in the Highlands of Scotland. Its egg is pale rufous, with dufky red fpots.

The Greenlanders catch the Ptarmigan by flipping a noofe over
its neck. Sometimes they kill it with ftones ; but now they commonly fhoot it. They eat the bird with feals fat, train-oil, and
berries, and efteem the repaft a great luxury. They make fhirts
of its plumage.

The Laplanders take thefe birds by forming a hedge of birchen-
boughs, and leave certain intervals, in which they hang fnares.

HUDSON's BAY PTARMIGAN.

Tetrao Albus.
Ripa Major, Schœf.
The White Partridge, Edw.
The White Grous, Penn. and Lath.

THE authors of the Britiſh Zoology juſtly blame Briſſon for claſſing the Ptarmigan with Edwards's White Partridge, ſince they are diſtinct ſpecies; for the latter is thrice as large as the Ptarmigan, and the colours of their ſummer garb are alſo very different, the White Partridge having broad ſpots of white and deep orange, and the Ptarmigan ſtreaks of a duſky brown on a light brown. The ſame authors admit, that in winter both birds are alike, almoſt entirely white. Edwards ſays, that the lateral quills of the tail are black even in winter, and only tipt with white; and yet he afterwards ſubjoins, that one of theſe which had been killed in that ſeaſon, and brought from Hudſon's-bay by Light, was of a ſnowy white, which ſtill more ſhews that in this ſpecies the colours of the plumage are variable.

The White Grous is of a middle ſize, between the partridge and the pheaſant, and its ſhape would reſemble that of the former, if its tail

were

were fomewhat fhorter. The one reprefented
in Edwards, Pl. LXXII. is a cock, fuch as it is in
fpring, when it begins to drop its winter's robe,
and feel the influence of the feafon of love; its
eyebrows are red and more prominent, and in
fhort, like thofe of the Red Grous; it has alfo
fmall white feathers round the eyes, and others
at the bottom of the bill, which cover the noftrils;
the two middle feathers are variegated like thofe
of the neck, the two fucceeding are white, and
all the reft blackifh, tipt with white, both in fummer and winter.

The livery of fummer extends only over the
upper part of the body; the belly continues always white, the feet and toes are entirely covered with feathers, or rather with white hairs; the
nails are lefs curved than ufual in birds *.

The White Grous refides the whole year in
Hudfon's-bay; it paffes the night in holes that
it makes in the fnow, which, in thefe arctic
countries, refembles fine fand. In the morning
it emerges from its retreat, and flies directly upwards, fhaking the fnow from off its wings. It
feeds in the morning and evening, and does not
feem to dread the fun, like the Ptarmigan of the
Alps; fince it fpends whole days expofed to his
rays, even in the middle of the day, when they

* We have feen two birds brought from Siberia under the name
of *Ptarmigans*, which were probably the fame fpecies with that of
Hudfon's Bay, and whofe nails were fo flat, that they refemble
more the nails of Apes than the claws of birds.

are

are moſt forcible. Edwards received this ſame bird from Norway, which appears to me to form the ſhade between the Ptarmigan and the Red Grous ; having the feet of the one, and the large eyebrows of the other. [A]

[A] Specific character of the White Grous, *Tetrao-Albus* :—" It " is orange, variegated with black ſtripes and white daſhes; its " toes ſhaggy ; its tail-quills black, and white at the tip ; the in- " termediate ones entirely white." The White Grous are amazingly numerous about Hudſon's Bay ; where they breed all along the coaſt, and lay about ten eggs, ſprinkled with black. In the beginning of October, they aſſemble in ſome hundreds, and live among the willows, whoſe tops they crop: Hence they are ſtyled *Willow Partridges.* In December they retire to the mountains to feed on cranberries : for, in that frightful climate, the cold is ſo intenſe, that the ſnow appears like fine powder, which in the depth of winter, is in a great meaſure ſwept by the winds from the up- lands, and carried into the plains. Theſe birds are generally tame as chickens ; if they chance to be unuſually ſhy, they may be ſoon hunted and worn out, till they ſink into their natural ſecurity. They are eſteemed excellent meat, and much ſought for by the ſervants of the Hudſon's Bay Company. They are commonly taken with nets of twine twenty feet ſquare ſet inclined, into which they are driven. Ten thouſand are often caught in the courſe of the winter.

FOREIGN BIRDS,

THAT ARE RELATED TO THE GROUS.

I.

The CANADA HAZEL GROUS.

La Gelinotte du Canada, Buff.
Tetrao Canadenfis, Linn. and Gmel.
Lagopus Freti Hudfonis, Klein.
The Black and Spotted Heath-cock, Edw.
The Spotted Grous, or *Wood Partridge,* Penn. and Lath.

IT would feem that the Hazel Grous of Canada, and the Hazel Grous of Hudfon's Bay, as defcribed by Briffon and defigned by Edwards, are the fame fpecies.

It is frequent through the whole year in the country bordering on Hudfon's Bay, and prefers the plains and low grounds; whereas, in another climate, the fame bird, fays Ellis, is found in the higheft tracts, and even on the fummit of mountains. In Canada, it is called the Partridge.

The male is fmaller than the common Hazel Grous; its eyebrows red; its noftrils covered with fmall black feathers; the wings fhort; the feet clothed below the tarfus; the toes and nails gray; the bill black. In general its colour is very dufky, and is brightened only by a few white

R 3   fpots

fpots round the eyes, on the flanks, and on fome other parts.

The female is fmaller than the male, and the colours of its plumage lighter and more varie-gated; in other refpects it is precifely alike.

Thefe birds feed on pine cones, juniper-berries, &c. They are numerous in the northern coun-tries of America, and are ftored up for winter's provifions; the froft preferves them from prutre-faction, and they are thawed in cold water, when they are to be ufed. [A]

[A] Specific character of the *Tetrao Canadenfis*—" Its tail-quills " are black, fulvous at the tip; two white dafhes at the eyes." By the Englifh fettlers at Hudfon's Bay it is called the *Wood Par-tridge*, becaufe it ufually lives among the pines. It is a very ftupid bird, often knocked down with a ftick, and commonly caught by the Indians with a noofe. In fummer, it lives on ber-ries; in winter, it crops the fhoots of the fpruce fir, which gives its flefh a difagreeable tafte. It is faid to lay only five eggs.

II.

The RUFFED HEATH-COCK,
Or, The LARGE HAZEL GROUS OF CANADA.

Tetrao Togatus, Linn. and Gmel.
Bonafa Major Canadenfis, Briff.
The Shoulder-knot Grous, Lath.

Though Briffon conceives this bird alfo as a diftinct fpecies from the ruffed Hazel Grous of Pennfylvania, it is highly probable that they are

really

really the fame : and to this fpecies muft we refer too the ruffed Heath-cock of Edwards. If we
confider that Edwards's figure was taken from a
living bird in love-feafon, and that Briffon's was
copied from a dead fubject ; if we make allowance for the liberties which are fuggefted by the
fancy of the defigner, we may difregard the
minute difparities.

It is rather larger than the ordinary Hazel
Grous, and like it, the wings are fhort, and the
feathers that cover the feet reach not to the
toes ; but it has neither the red eyebrows, nor
the ring of that colour which encircles the eyes.
What diftinguifh it, are the two tufts of feathers
which rife from the upper part of the breaft,
one on each fide, and project beyond the reft,
and bend downwards ; the feathers which form
thefe are of a fine black, the edges beaming with
different reflections of gold green. The bird can
expand at pleafure thefe falfe wings, which when
clofed fall on both fides on the upper part of the
true ; the bill, toes, and nails, are of a reddifh
brown.

This bird is, according to Edwards, very common in Maryland and Pennfylvania, where it is
called the *pheafant*. But its inftincts and habits
are much nearer thofe of the Grous. It is of a
middle fize between that of the pheafant and the
partridge ; its feet are feathered, and its toes indented on the edges like thofe of the Grous ; its
bill is fimilar to that of a common cock ; its

R 4　　　　　　　noftrils

noftrils are fhaded with fmall feathers, which
rife from the bottom of the bill, and point forward; the whole upper part of the body, including the head, the tail, and the wings, are
mailed with different brown colours, more or
lefs brightened with the mixture of orange and
black; the throat is of a brilliant orange, though
rather deep; the ftomach, the belly and the
thighs, are marked with black fpots, in the fhape
of a crefcent, and ftrewed with regularity on a
white ground : it is furnifhed with long feathers
round the head and neck, which it can erect at
will, and form a creft or ruff, and this it generally does in the feafon of its amours; it alfo
fpreads the tail-quills like a fan, inflates its craw,
trails its wings, and ruftles with a whirring noife
like a turkey-cock; it fummons its females alfo
by a very odd fort of clapping the wings, which
is fo loud as to be heard at half-a-mile's diftance
in calm weather *. It takes this kind of exercife in fpring and autumn, which are the feafons
of breeding, and repeats it every day at ftated
hours, viz. at nine o'clock in the morning, and
four o'clock in the afternoon, and this always
fitting on a dead trunk. At firft, it ftrikes flowly, allowing an interval of two feconds between
each beat; but it gradually quickens the ftrokes,
which at laft become fo rapid as to appear a

* Mr. Bartram fays, that the people of Pennfylvania call this
the *thumping* of the ruffed Grous.

continued

continued found, refembling the noife of a
drum, or, according to fome, the muttering
of diftant thunder. This noife lafts about a
minute, and, after a repofe of feven or eight
minutes, it again renews and paffes through the
fame gradations. Such is the call which invites
the female to the feaft of love; but what an-
nounces a future generation, is often the fignal
for the deftruction of the prefent. The fportfman, led by the noife, approaches the bird unper-
ceived, and when the male is diffolved in con-
vulfions of pleafure, he takes the fatal aim. If
the bird however obferve the perfon, it ftops its
motions, and flies off three or four hundred
paces.—Thefe are really the inftincts and habits
of the European Grous, though the fingularities
are rather heightened.

The common food of thofe in Pennfylvania
is grain, fruits, wild grapes, and above all ivy
berries, which is the more extraordinary, as
thefe prove fatal to other animals.

They hatch only twice a year, probably in
fpring and autumn, which are the two feafons
when the male beats his wings. They make
their nefts on the ground with leaves, or by the
fide of a fallen trunk, or at the foot of a tree;
all which habits indicate a heavy bird. They
lay from twelve to fixteen eggs, and fit about
three weeks. The mother has the fafety of her
young much at heart; fhe rifks every thing in
their defence, and expofes herfelf to all the

dangers

dangers that menace their deſtruction. The
tender brood are themſelves dexterous in ſearch-
ing for a concealment beneath the leaves. But
all theſe precautions are inſufficient to elude the
dreaded aſſaults of the birds of prey. The little
family continues united, till the glow of the following ſpring inſpires new appetites, and diſ-
perſes its members.

Theſe birds are exceedingly wild, and can
never be tamed. If they are hatched under
common hens, they fly almoſt as ſoon as they
burſt from the ſhell, and hide themſelves in the
foreſts. The fleſh is white, and an excellent meat; and may not this be the reaſon
why the rapacious birds chaſe them with ſuch
perſeverance? We have already mentioned
the conjecture in treating of the European
Grous; if it were confirmed by a ſufficient number of obſervations, we might infer that vora-
city does not always exclude predilection, but
that the birds of prey have nearly the ſame taſte
as man: and this would afford another analogy
between thoſe two ſpecies. [A]

[A] Specific character of the Ruffed Grous, *Tetrao-Umbellus*:
" It has a ruff about its neck." Its fleſh is lean, dry, cloſe, and
exceedingly white; yet if well cooked, it is excellent food. The
bird builds its neſt on dry ground, and hatches nine young. The
mother clucks to her chickens, and gathers them under her
wings.

III. The

III.

The LONG-TAILED GROUS.

Tetrao Phafianellus, Lin. and Gmel.
The Sharp-tailed Grous, Penn.

The American bird, which may be called the Long-tailed Hazel Grous, defigned and defcribed by Edwards under the name of the *Hudfon's-Bay Heath-Cock*, or *Grous*, but which appears to me to be more related to the Hazel Grous. The individual reprefented in Edwards, Plate CXVII, is a female, with the fize, colour, and long tail of the pheafant; the plumage of the male is of a deeper fhining brown, with various reflections near the neck: and he ftands very erect, with a bold afpect; differences which are invariable between the male and female in all birds of this kind. Edwards did not venture to give red eye-brows to this female, becaufe he only faw a ftuffed fpecimen, in which that character was not fufficiently diftinct; the legs were rough, the toes indented on the edges, and the hind toe very fhort.

At Hudfon's Bay, this bird is called a *pheafant*. The long tail, indeed, forms a fort of fhade between the hazel grous and the pheafants. The two middle quills of the tail project

two

two inches farther than the two following on either fide, and thus gradually fhorten. Thefe birds are alfo found in Virginia, in the woods and the unfrequented parts. [A]

[A] Specific charaᵭer of the Long-tailed Grous, *Tetrao Pha-fianellus* :—" Its tail is wedge-fhaped ; its head, its neck, and the " upper fide of its body, are brick coloured, ftriped with black." In Hudfon's Bay, it lives among the larch bufhes : feeds on berries in fummer, and on the buds of larch and birch in winter. It lays from nine to thirteen eggs. The cock has a very fhrill fort of a crow, not very loud. When difturbed, or on wing, he repeats the found *cuck, cuck,* and cracks the feathers of his tail. The flefh of thefe birds is gray, fat, and juicy.

M

THE CRESTED PEACOCK

The PEACOCK*.

Le Paon, Buff.
Pavo Criſtatus, Linn. and Gmel.
The Creſted Peacock, Lath.

IF empire belonged to beauty and not to
ſtrength, the Peacock would undoubtedly be
king of the birds; for upon none of them has
Nature poured her treaſures with ſuch profuſion.
Dignity of appearance, nobleneſs of demeanour,
elegance of form, ſweetneſs and delicacy of proportions, whatever marks diſtinction and com-
mands reſpect, have been beſtowed. A light
waving tuft, painted with the richeſt colours,
adorns its head, and raiſes without oppreſſing
it. Its matchleſs plumage ſeems to combine all
that delights the eye in the ſoft delicate tints of
the fineſt flowers; all that dazzles it in the
ſparkling luſtre of the gems; and all that aſto-
niſhes it in the grand diſplay of the rainbow.
But not only has Nature united, in the plumage

---

* In Greek, Tαως, or Tαων, perhaps from τεινω, to ſtretch, on
account of the length of its tail: in the Æolian dialect it was pro-
nounced Παων; and hence the Latin *Pavo*, and its names in the
modern languages: in Italian, *Pavone*; in Spaniſh, *Pavon*;
in French, *Paon*; in German, *Pfau*; in Poliſh, *Paw*; and in
Swediſh, *Pao-fogel*.

of

of the Peacock, to form a mafter-piece of magnificence, all the colours of heaven and earth;
fhe has felected, mingled, fhaded, melted them
with her inimitable pencil, and formed an un-
rivalled picture, where they derive from their
mixture and their contraft new brilliancy, and
effects of light fo fublime, that our art can neither imitate nor defcribe them.

Such appears the plumage of the Peacock,
when at eafe he faunters alone in a fine vernal
day. But if a female is prefented fuddenly to
his view; if the fires of love, joined to the
fecret influence of the feafon, roufe him from
his tranquillity, and infpire him with new
ardour and new defires; his beauties open and
expand, his eyes become animated and expreffive, his tuft flutters on his head, and expreffes
the warmth that ftirs within; the long feathers
of the tail, rifing, difplay their dazzling rich-
nefs; the head and neck bending nobly backwards, trace their fhadow gracefully on that
fhining ground, where the fun-beams play in a
thoufand ways, continually extinguifhed and
renewed, and feem to lend new luftre, more
delicious and more enchanting; new colours,
more variegated and more harmonious; each
movement of the bird produces new fhades,
numberlefs clufters of waving, fugitive reflections, which ever vary and ever pleafe.

It is then that the Peacock feems to fpread out
all his beauties, only to delight his female, who,
though

though denied the rich attire, is captivated with its difplay; the livelinefs which the ardor of love mingles with his geftures, adds new grace to his movements, which are naturally noble and dignified, and which, at this time, are accompanied with a ftrong hollow murmur expreffive of defire *.

But this brilliant plumage, which furpaffes the glow of the richeft flowers, like them alfo is fubject to decay; and each year, the Peacock fheds his honours †. As if afhamed at the lofs of his attire, he avoids being feen in this humiliating condition, and conceals himfelf in the darkeft retreats, till a new fpring reftores his wonted ornaments, and again introduces him to receive the homage paid to beauty; for it is pretended, that he is really fenfible to admiration, and that a foothing and attentive gaze is the moft certain means to engage him to difplay his decorations; but the look of indifference chills his vivacity and makes him clofe his treafures.

Though the Peacock has been long naturalized in Europe, it is not a native of this quarter of the globe. The Eaft Indies, the climate that produces the fapphire, the ruby, and the topaz, muft be confidered as the origi-

---

* " Running forward with a creeking noife." PALLADIUS.

† It lofes its feathers with the firft fall of the leaves, and recovers them again when the buds burft forth.

ARISTOTLE, *Hift. An.*

4

nal

nal country of the moſt beautiful of birds.
Thence it paſſed into the weſtern parts of Aſia,
where, according to the expreſs teſtimony of
Theophraſtus, quoted by Pliny, it had been
introduced from abroad *. But it does not appear to have been carried thither from the
eaſtern part of Aſia, or China; for travellers
agree, that though very common in the Eaſt
Indies, it is not indigenous in China, which at
leaſt proves it to be a rare bird in that country †.

Ælian informs us, that Greece received this
beautiful bird from the Barbarians ‡; who
muſt have been the people of India, ſince
Alexander, who traverſed Aſia, and was well
acquainted with Greece, firſt met with the Peacock in that country §: and beſides, in no
region of the globe is the tribe ſo numerous as
in that oriental clime. Mandeſlo and Thevenot
ſaw them in profuſion in the province of Gu-
zarat; Tavernier, in every part of India, but
particularly in the territories of Baroche, Cambaya, and Broudra; Francis Pyrard, in the
vicinity of Calicut; the Dutch, on the Mala-
bar coaſt; Lintſcot, in the iſland of Ceylon;
the Author of the Second Voyage to Siam, in
the foreſts on the frontiers of that kingdom;
on the ſide of Cambogia, and near the banks

---

* " Theophraſtus relates, that even in Aſia the pigeons and
" peacocks are of foreign extraction." PLIN. lib. x. 29.

† Navarette, *Deſcription de la Chine.*

‡ *Hiſt. Anim.* lib. v. 21.   § Id. ibid.

of the river Meinam; Gentil, at Iva; Gemelli
Carreri, in the Calamian iflands, lying between
the Philippines and Borneo: if to thefe authorities we add, that in all thefe countries the
Peacocks live in the wild ftate, and that they
are no where elfe fo large*, or fo prolific †,
we cannot hefitate to conclude that the Eaft
Indies is their native abode. That beautiful
bird muft owe its birth to the luxurious climate
where Nature lavifhly pours her riches; where
gold, and pearls, and gems, and precious
ftones, are fcattered with profufion. This opinion is countenanced by Holy Writ; Peacocks
are enumerated among the valuable and rare
commodities that were every three years imported by Solomon's fleet; which being fitted
out in the Red Sea, and not being able to venture
at a diftance from the fhore, muft obvioufly have
drawn its riches either from India, or the
eaftern coaft of Africa. Nor is it probable that
the latter was the place that furnifhed thefe
luxuries; for no traveller has ever feen wild
Peacocks in Africa, or the adjacent iflands;
except at St. Helena, where Admiral Verhowen
fhot fome that could not be caught. But it is
not probable that Solomon's fleet could fail
every three years to Madeira, without a mari-

---

* " The largeft Peacocks are found in India."
ÆLIAN, lib. xxvi. 2.
† Peter Martyr, *de Rebus Oceani,* fays, that in India the Pea-
hens lay from twenty to thirty eggs.

ner's

ner's compafs; where, befides, they could obtain neither gold, nor filver, nor ivory, nor
fcarce any thing which they might want. I
fhould even imagine that in this ifland, which is
above three hundred leagues from the continent,
there were no Peacocks in Solomon's time, and
that thofe found there by the Dutch, had been
left by the Portuguefe, and had multiplied exceedingly in the wild ftate; efpecially as it is
faid that no venomous creature or voracious
animal exifts in St. Helena.

Nor can we doubt that the Peacocks which
Kolben faw at the Cape of Good Hope, and
which, he fays, are exactly like thofe of Europe,
though the figure that he gives is widely different *, had the fame origin with thofe at St.
Helena, and had been carried thither in fome
of thofe European fhips which are continually
vifiting that coaft.

The fame may be faid of thofe feen by
travellers in the kingdom of Congo †, with
the turkies, which undoubtedly are not natives of Africa; and of thofe alfo that are
found on the confines of Angola, in a wood
inclofed by a wall, where they are bred for the
king of the country ‡. This conjecture is cor-

* *Hift. Gen. des Voyages*, tome v. pl. 24.

† *Voyage de P. Van-Brœk*, in the *Recueil des Voyages qui
ont fervi a l'etabliffement de la Compagnie des Indes*, tome iv.
p. 321.

‡ *Relation de Pigafetta*, p. 92.

roborated

roborated by the teftimony of Bofman, who exprefsly mentions that there are no Peacocks on the Gold Coaft, and that the bird taken by de Foquembrog and others for a Peacock, is quite different, and called *Kroon vogel* *.

Befides, the term *African Peacock*, beftowed by moft travellers on the *Demoifelle of Numidia* †, is a direct proof that Africa is not the natal region of the Peacock. If they were anciently feen in Lybia, as Euftathius relates, they were certainly tranfported from India to that country, which is the part of Africa next to Paleftine ; nor does it appear that they were naturalized in that country, or multiplied faft, fince fevere laws were paffed againft killing or wounding them ‡.

We may therefore prefume, that Solomon's fleet did not import thefe rarities from the African coaft, but from the fhores of Afia, where they abound, living in a ftate of nature, and multiplying without the affiftance of man; and where they are larger and more prolific than in other countries, as is the cafe with all animals in their congenial climate.

From India they migrated into the weftern part of Afia. Accordingly we learn from Diodorus Siculus, that they abounded in Ba-

* *Voyage de Guinée*, Lettre xv.
† Labat.—*Voyage de M. de Genes au detroit de Magellan.*
‡ Aldrovandus.

bylon.

bylon. In Media alfo they were bred in fuch numbers, that the bird was called *Avis Medica* *. Philoftratus fpeaks of thofe of Phafis, which had a blue creft †, and travellers have feen fome of that kind in Perfia ‡.

From Afia they were tranfported into Greece, where at firft they were fo rare as to be exhibited in Athens for thirty years, at the monthly feftivals, as an object of curiofity, which drew crouds of fpectators from the neighbouring towns §. We cannot fix the date of this event; but we are certain that it was after the return of Alexander from India, and we know that he firft ftopped at the ifland of Samos. The conqueror was fo delighted with the rich plumage of the Peacocks, that he enacted fevere penalties againft killing them. But it is very probable that foon after his time, and even before the clofe of his reign, they were become common; for we learn from the poet Ariftophanes, who was contemporary with that hero and furvived him, that a fingle pair brought into Greece had multiplied fo rapidly, that they were as numerous as quails; and

---

* Aldrovandus.   † Idem.

‡ Thevenot, *Voyage du Levant*.

§ " The Peacock was at Athens fhewn for a ftated price to " both men and women, who were admitted to the fpectacle at " the feafts of the new moon. Confiderable fums were thus col- " lected; and many, through curiofity, came from Lacedæmon " and Theffaly." ÆLIAN, *Hift. Anim.* vol. 2.

besides,

befides, Ariftotle, who outlived his pupil only two years, fpeaks in feveral parts of his work of Peacocks as well-known birds.

Secondly, That the ifle of Samos was the firft ftation of Alexander on his return from India, is probable from its proximity to Afia; and is befides proved by the exprefs teftimony of Menodotus *. Some indeed have given a forced interpretation of this paffage, and refting on the authority of fome very ancient medals of Samos, in which Juno is reprefented with a Peacock at her feet †, have pretended that Samos was the primitive abode of that bird, from whence it has been difperfed to the eaft and the weft. But if we examine the words of Menodotus, we fhall find that they mean no more than that Samos was the firft part in Europe where the Peacocks were bred; in the fame manner as the Pintadoes, which are well known to be African birds, were feen in Æolia or Ætolia, before they were introduced into the reft of Greece; and efpecially as the climate of Samos is particularly fuited to them ‡, and they lived there in the ftate

---

* " There are the Peacocks facred to Juno, they being firft reared " in Samos, and thence carried into other countries, as the cocks " from Perfis, and the Meleagrides from Æolia (or Ætolia)."

ATHENÆUS.

† Some of thefe are ftill to be feen, and even medallions which reprefent the temple of Samos, with Juno and her Peacocks. TOURNEFORT's *Voyage to the Levant.*

‡ Foreign flocks of Peacocks are faid to fubfift wild on the ifland of Samos in the grove of Juno.

VARRO, *de Re Ruftica,* lib. iii. 6.

of

of nature; and as Aulus Gellius confidered thofe of that ifland as the moft beautiful of all *.

Thefe reafons are more than fufficient to account for the epithet of Samian bird, which fome authors have beftowed on the Peacock; but the term can no longer be applied, fince Tournefort never mentions the Peacock in his defcription of that ifland, and fays that it is full of partridges, woodcocks, thrufhes, wild-pigeons, turtles, becafigoes, and excellent poultry; and it is not probable that Tournefort would include fo diftinguifhed a bird in the generic term *poultry*.

After the Peacock was tranfplanted from Afia into Greece, it found its way into the fouth of Europe, and gradually was introduced into France, Germany, Switzerland †, and as far as Sweden, where indeed they are very rare, and require great attention ‡, and even fuffer an alteration in their plumage.

Laftly, The Europeans, who by the extent of their commerce and navigation connect the whole inhabited world, have fpread them along the African coafts, and adjacent iflands; and

---

* *Noct. Attic.* l. vii. c. 16.

† The Swifs are the only people who have endeavoured to extirpate this beautiful fpecies of bird, and with as much pains as other nations have beftowed in rearing them. The reafon is fomewhat whimfical; the creft of the Dukes of Auftria, againft whom they had revolted, was a Peacock's tail.

‡ Linnæus

after-

afterwards introduced them into Mexico, Peru, and fome of the Antilles *, as St. Domingo and Jamaica, where they now are numerous †, though there were none prior to the difcovery of America. The Peacock is a heavy bird, as the ancients well remarked ‡; the fhortnefs of its wings, and the length of its tail, check its aërial courfe; and as it with difficulty fubfifts in a northern climate §, it could never migrate into the new world.

The Peacock has fcarcely lefs ardour for the female, or contends with lefs obftinacy, than the common cock ‖. His paffions muft even be more fiery, if it be true, that when he has only one or two hens, he teazes and fatigues them, and even induces fterility and difturbs the work of generation, by his immoderate ufe of venery. In this cafe, the eggs are ejected from the oviduct before they have time to ripen ¶; and, for this reafon, he ought to be allowed five or fix females **; whereas, when the ordinary cock,

* Hiftoire des Incas.

† Charlevoix's Hiftory of St. Domingo, and Ray's Synopfis of Birds.

‡ " They can neither foar high, nor fly to great diftances."
COLUMELLA.

§ " They live fometimes with us, efpecially in the aviaries of " the great, but require attention." LINNÆUS.

‖ Columella de Re Ruftica, lib. viii. 11. ¶ Id. Ibid.

** I here give the opinion of the ancients; for intelligent perfons whom I have confulted, and who have reared Peacocks in Burgundy, affure me from experience, that the males never fight, and that hey require each only one or two females at moft. But perhaps this coolnefs of paffion is owing to the nature of the climate.

who

who can fatisfy the wants of fifteen or twenty hens, is reduced to one, he makes her the mother of a numerous brood.

The pea-hens are also of an amorous mould, and when deprived of the males, they toy with each other, and welter in the duft; but the eggs which they lay are then void of the principle of life. This happens commonly in the fpring, when the return of foft and genial warmth awakens nature from her torpor, and gives a new ftimulus to the appetite, which prompts every animated being to reproduce its fpecies. Hence perhaps the reafon why fuch eggs were termed zephyrian *(ova zephyria)*, not becaufe the gentle zephyrs were imagined capable of impregnating them, but becaufe the vernal feafon is fanned by light airs, and even depicted by the zephyrs *.

I could eafily believe that the fight of the male ftrutting round them, difplaying his tail, and fhewing every expreffion of defire, would ftill more excite them, and make them lay more of thefe addle eggs; but I will never be perfuaded that the careffes, diftant geftures, and light flutterings, would effect a real fecundation, without the more intimate union, and the more vigorous compreffions of the male. And the pea-hens which fome have fancied to be im-

---

* *Wind-eggs* is their common name in Englifh; becaufe they want the outer fhell, and are flaccid, as if inflated with air. Perhaps this was alfo the reafon of the ancient epithet Zephyrian.

pregnated

pregnated by the influence of love glances, muſt
have been covered before, though unobſerved *.

Theſe birds, according to Ariſtotle, attain their
full vigour in three years †. Collumella ‡ is of
the ſame opinion, and Pliny repeats the words
of Ariſtotle, with ſome ſlight alterations §. Varro
fixes the period at two years ‖ ; and people who
are well acquainted with theſe birds inform me,
that in our climate the female begins to lay at
the end of the year, though the eggs are then
certainly addle. But almoſt all agree that the
age of three years is the term when the Peacock has acquired his full growth, and is fit to
perform the office of the male; and that the
power of procreating is announced by a new and
ſplendid production : this is the long and beautiful feathers of the tail, which they diſplay, as
they ſtrut and expand their fan ¶ ; the ſurplus
nouriſhment being no longer directed to the
growth of the individual, is ſpent on the reproduction of the ſpecies.

The ſpring is the ſeaſon when theſe birds ſeek
to couple ** : and if we would forward the
union, we muſt, according to Columella's direc-

* Belon makes the ſame remark.　† Hiſt. Anim. lib. vi. 9.
‡ De Re Ruſtica, lib. viii. 11.
§ " The firſt year it lays one or two eggs, the ſecond four or
" five, and in the following years not more than twelve." Lib.
x. 59.
‖ Lib. iii. 6.　¶ Pliny, lib. x. 20.
** " About the ides of February, before the month of March."
　　　　　　　　　　　　　　　　COLUMELLA.
　　　　　　　　　　　　　　　　　tion,

tion, give them, every five days, in the morning while fasting, beans slightly roasted *.

The female lays her eggs soon after fecundation; she does not exclude one every day, but only once in three or four days, and according to Ariftotle she has but one hatch in the year, which confifts in the firft of eight eggs, and in the following years of twelve. But this muft be underftood of thofe pea-hens that both lay their eggs and rear their young; for if the eggs be removed as faft as they are laid, and are placed under a common hen †, they will, according to Columella, have three hatches in the courfe of the year; the firft of five eggs, the fecond of four, and the third of two or three. It would feem that in this country they are not fo prolific, fince they lay fcarcely four or five eggs in the year. On the other hand, they appear to be far more prolific in India, where, according to Peter Martyr, they lay twenty or thirty, as I have already noticed. The temperature of a climate has a mighty influence on whatever relates to genera-

---

* COLUMELLA.

† Ariftotle fays, that an ordinary hen cannot hatch more than two pea-hens' eggs; but Columella allows five of thefe eggs in addition to four common eggs. He advifes to remove the eggs the tenth day, and fubftitute an equal number of the fame kind recently laid, in order that they may be hatched along with the pea-hen's eggs, which require ten days longer incubation. Laftly, he direfts that thefe be turned every day, if the fitter be unable to do it on account of their bulk, which it is eafy to difcover by marking the eggs on one fide.

tion,

tion, and this is the key to thofe apparent contradictions which are found between the writings of the ancients and our own obfervations. In a warm country, the males are more ardent, fight with each other, require more females, and thefe lay a greater number of eggs; but in a cold country the latter are not fo prolific, and the former are calm and indifferent.

If the pea-hen be fuffered to follow the bent of inftinct, fhe will lay her eggs in a fecret retired fpot ; the eggs are white, and fpeckled like thofe of the turkey-hen, and nearly of the fame fize. It is afferted that fhe is very apt to lay in the night, or rather carelefsly drop the eggs from the rooft on which fhe is perched; and for this reafon, it is advifed to fpread ftraw underneath, to prevent their being broken by the fall *.

During the whole time of incubation, the pea-hen anxioufly fhuns the male, and is particularly careful to conceal her track, when fhe returns from the neft : for in this fpecies, as in the gallinaceous tribe and many others †, the male burning with luft, and faithlefs to the intentions of nature, is more earneft in the purfuit of pleafure, than folicitous about the multiplication of the race. If he difcovers his mate fitting on her eggs, he breaks them; probably to remove an obftacle to the gratification of his paf-

* Columella, lib. viii. 11. † Ariftotle Hift. Anim. lib. vi. 9.

fions.

fions. Some have imagined that it was from the defire of covering them himfelf*, which would be a very different motive. Natural hiftory will continually be clouded with uncertainties; to remove them, we ought to obferve every thing ourfelves; but who is able for the tafk?

The pea-hen fits from twenty-feven to thirty days, more or lefs, according to the temperature of the climate, and the warmth of the feafon †. During that time, a fufficient fupply of food ought to be fet within their reach, that they may not be obliged to ftray in fearch of fubfiftence, and allow their eggs to cool; and care muft be taken not to teaze or difturb them in their neft; for if they perceive that they are difcovered, they will be filled with difquietude, abandon their eggs, and begin to make a fecond hatch, which is not likely to fucceed, becaufe of the latenefs of the feafon.

It is faid that the pea-hen never hatches all her eggs at once, but as foon as a few chickens emerge, fhe leaves the neft to lead them about. In this cafe, the eggs that are left fhould be fet under another hen, or placed in a ftove for incubation ‡.

Ælian tells us, that the pea-hen does not fit conftantly on her eggs, but fometimes leaves

---

* Aldrovandus.   † Ariftotle, lib. vi. 9. and Pliny, lib. x. 59.
‡ Maifon Ruftique, tome i. p. 138.

them

them two days together, which fufpends the
progrefs of incubation. But I fhould imagine
that there is fome miftake in the text of Ælian,
which refers to the hatching, what Ariftotle and
Pliny mention with regard to the laying, which
is really liable to interruptions of two or three
days; whereas fuch interruptions in the fitting
feem to be inconfiftent with the law of nature
obferved by all the known fpecies of birds, unlefs when the heat of the climate approaching
that of incubation difpenfes with it as unneceffary *.

After the young are hatched, they ought to
be left under the mother for twenty-four hours,
and then removed to the coop † ; Frifch advifes
them not to be reftored to their dam till fome
days after.

Their firft food muft be barley-meal, foaked
in wine ; wheat fteeped in water ; or even pap
boiled, and allowed to cool. Afterwards they
may have frefh curd, from which the whey
is well preffed, mixed with chopped leeks, and
even grafshoppers, of which they are very fond,
but the legs muft be previoufly removed from
thefe infects ‡. When they are fix months old,
they will eat wheat, barley, the dregs of cyder
and perry, and even crop the tender grafs ; but
that fort of nourifhment is not fufficient, though
Athenæus reprefents them as *graminivorous*.

---

* As in the cafe of the Oftrich.   † Columella, lib. viii. 11.
‡ Columella, lib. viii. 11.

It is obferved that on the firft days after hatching, the mother never leads her young to the
ordinary neft, or even fits with them twice in
the fame place ; and as they are delicate, and
cannot mount on the trees, they are expofed to
many accidents.   At this time therefore we
ought to watch them clofely, and difcover where
the mother reforts, and put the brood in a coop,
or in the field in a patch inclofed with hurdles,
&c. *

Till they grow ftout, the young Peacocks
trail their wings †, and make no ufe of them.
In their early effays to fly, the mother takes
them every evening one after another on her
back, and carries them to the branch on which
they are to pafs the night.   In the morning, fhe
defcends before them from the tree, and en-
courages them by her example to truft themfelves to their flender pinions ‡.

A pea-hen, or even a common hen, can breed
twenty-five young Peacocks, according to Columella ; but only fifteen, according to Palladius : and this laft number is even too great for
cold countries, where they muft be warmed
from time to time, and fheltered under the
mother's wing.

It is faid that the common hen, when fhe fees
a hatch of young Peacocks, is fo pleafed with
their beauty, that fhe grows difgufted with her

---

* Maifon Ruftique, tome i. p. 138.

† Belon.   ‡ Maifon Ruftique, tome i. p. 139.

own

own chickens, and attaches herself to the strangers *. I mention this circumstance not as a fact that is ascertained, but as one that deserves to be inquired into.

As the young Peacocks grow strong, they begin to fight, (especially in warm countries,) and for this reason the ancients, who seem to have bestowed more attention than we in training these birds †, kept them in small separate huts ‡. But the best places for breeding were, according to them, the islets, which are so numerous on the Italian coasts §; for instance, that of Planasia, belonging to the Pisans ‖. Such a spot indeed allowed them to follow freely the bent of nature, without danger of escaping, since they are unable to fly to a distance, and cannot swim; and at the same time they had nothing to apprehend from rapacious animals, which were entirely extirpated from the little island. They lived there at ease, without constraint, and without disquietude; they thrived better, and (what was not overlooked by the Romans) their flesh acquired a finer relish; and to have them under their eye, and to examine whether their numbers increased or diminished, they accustomed them every day at a stated hour, on the display of a certain signal,

* Columella, lib. viii. 11.　　† Id. ibid.
‡ Varro, *De Re Rustica*, lib. iii. 6.
§ Columella, *loco citato*.　　‖ Varro, *loco citato*.

to

to come round the houfe, and they threw a few handfuls of grain to draw them together *.

When the brood are a month old, or a little more, the creft begins to fhoot, and then they are fubject to ficknefs, as the young turkies in fimilar circumftances. At this time the parent cock adopts them as his offspring; for before the growth of the creft, he drives them away as fuppofititious †. They ought not however to be trufted with the old ones before the age of feven months, and they muft be accuftomed to perch on the rooft, that they may not fuffer from lying on the ground, on account of the cold damps ‡.

The creft confifts of fmall feathers, of which the fhaft is not furnifhed with webs, but befet with little flender detached threads; the top is formed by a bunch of ordinary feathers united together, and painted with the richeft colours.

The number of thefe fmall feathers is variable; I have counted twenty-five in a male, and thirty in a female; but I have not examined enough to decide accurately.

The creft is not an inverted cone, as might be fuppofed; its bafe, which is uppermoft, forms a very extenfive ellipfe, whofe greater axis is in the direction of the head; all the feathers that compofe it, have a particular and perceptible

---

* Columella, *loco citato.*
† Palladius, *De Re Ruftica*, lib. i. 28.   ‡ Columella.

8   motion,

motion, by which they approach each other, or recede, at will, and alfo a general motion, by which the whole creft is fometimes erected, fometimes reclined.

The waving fummits of this creft, as well as all the reft of the plumage, are decorated with much more fplendid colours in the male than in the female. Befides this circumftance, the cock is difcriminated from the hen when three months old, by a little yellow which appears on the tip of the wing; he is afterwards diftinguifhed by his fize, by the fpur on each leg, by the length of his tail, and the power of expanding it like a fan. Willughby fancies that the Peacock fhares that remarkable property with the turkey alone; but in the courfe of this hiftory we have feen that it belongs alfo to fome grous, to fome pigeons, &c.

The tail-feathers, or rather thofe long coverts that are inferted in the back near the rump, are on a great fcale what thofe of the creft are on a fmall one. The fhaft is equally furnifhed from its origin to its extremity, with parted filaments of a varying colour, and it ends in a flat vane, decorated with what is called the *eye*, or the *mirror*. This is a brilliant fpot, enamelled with the moft enchanting colours; yellow, gilded with many fhades, green running into blue and bright violet, according to the different pofitions, and the whole receives additional luftre from the colour of the centre, which is a fine velvet black.

The two feathers in the middle are each four feet and a half long, and extend beyond the reft, the others gradually diminifhing as they approach the fides. The creft is permanent, but the tail is caft every year, either entirely or in part, about the end of July, and fhoots again in the fpring; during which interval the bird is difpirited and feeks retirement.

The predominant colour of the head, throat, neck, and breaft is blue, with different reflections of violet, yellow, and lucid green; and by means of thefe waving fhades, nature can fpread a greater variety of colouring on the fame fpace.

On each fide of the head, there is a protuberance formed by fmall feathers, which cover the perforation of the external ear.

Peacocks feem to toy with each other by the bill; but on examining them clofely, I find that they fcratch the head, which is fubject to a very nimble fort of lice. Thefe may be feen running over the white fkin that encircles the eyes, which muft occafion an uneafy feeling. Accordingly, the birds remain very tame and feem pleafed when another fcratches them.

Thefe birds affume the rule in the yard, and will not fuffer the other poultry to feed till they have fatisfied their hunger. They eat nearly the fame way with the gallinaceous tribe, laying hold of the grain by the point of the bill, and fwallowing it whole.

When

When they drink, they plunge their bill into the water, and make five or fix quick motions with the lower jaw; then raifing their head and holding it horizontal, they fwallow the water, with which their mouth is filled, and without moving the bill.

Their food is received into the *æfophagus*, where a little above the anterior orifice of the ftomach, is placed a glandulous fwelling filled with fmall tubes, which pour out much limpid liquor.

The ftomach is clothed on the outfide with a great number of mufcular fibres.

In one of thefe birds, which was diffected by Gafpar Bartholin, there were two biliary ducts; but he found only one pancreatic duct, though there are generally two in the feathered tribes.

The *cæcum* was double, and pointing from behind forwards; its length was equal to that of all the other inteftines together, and was more capacious *.

The rump is very thick, becaufe in it are inferted all the mufcles deftined to elevate and expand the tail.

The excrements are commonly figured, and mixed with a little of that white matter which is common to the gallinaceous tribes, and many other birds.

I am informed that they fleep, fometimes hiding their head under their wing, fometimes covering their neck, and leaving the bill expofed.

* Acta Hafnienfia, 1673.

Peacocks

Peacocks love cleanlinefs, and for this reafon
they are at pains to hide their excrements; not
becaufe they are loth that men fhould derive
any benefit from the dung *, which it is faid is
good for fore eyes, for manure, &c. but doubt-
lefs they are not well acquainted with all thefe
properties.

Though they cannot fly much, they are fond
of climbing. They generally pafs the night on
the roofs of houfes, where they do a great deal
of mifchief, and on the loftieft trees. From
thefe elevated ftations, they often fcream; and
their cry is univerfally allowed to be difagreeable,
perhaps becaufe it difturbs our fleep, and from
which it is pretended that their name is formed in
all languages †.

It is faid that the female has only one note,
which fhe feldom utters except in the fpring,
while the male has three. For my own part, I
can only diftinguifh two tones; the one flat,
like that of the hautboy, the other fharp, exactly the octave of the former, which refembles
more the fhrill notes of the trumpet; and I con-
fefs that my ear is not hurt by thefe founds, any
more than my eye by the fhape of their legs:
and we apply to the Peacocks our falfe reafon-

* " Fimum fuum reforbere traduntur, invidentes hominum uti-
" litatibus." PLIN. lib. xxix. 6. Hence the Peacock is faid to
be envious.

† Volucres pleræque a fuis vocibus appellatæ, ut hæ . . . . .
Upupa, cuculus, ulula, pavo. VARRO de Linguâ Latinâ.

ings

ings and even our vices, when we fuppofe that their cry is only a groan extorted by their vanity, as often as they view the clumfinefs of their feet.

Theophraſtus maintains, that their cries if often reiterated, forebode rain ; others, that they foretell it when they fcramble higher than ordinary *. Others allege that thefe cries forebode the death of a neighbour ; and laftly, others relate that thefe birds always wear under the wing a bit of the root of flax, as an amulet to preferve them from witchcraft †. Whatever is much fpoken of, is made a fubject of filly fables.

Befides the different cries which I have mentioned, the male and female emit a certain dull found, or fmothered cracking, which feems to be formed internally, and which they often repeat, whether they are difturbed or in a ftate of tranquillity and eafe.

Pliny fays, that a fympathy has been obferved between the pigeons and the Peacocks ‡; and Clearchus tells us of one of the latter which grew fo much attached to a young woman, that, having witneffed her death, it could not furvive the fhock §. But a more natural and better founded friendſhip is obferved between the turkey and Peacock. Thefe two birds are of the number that raife and difplay their tail ; a circumftance which implies many common pro-

* De Naturâ Rerum. † Ælian. Hift. Anim. lib. xi. 8.
‡ Plin. Hift. Nat. lib. x. 20. § Athenæus, lib. xiii. 20.

perties.

perties. Accordingly, they agree better than with the other fowls. It is even faid that a Peacock has been feen to copulate with a turkey-hen *; which would fhew a great analogy between the two fpecies.

The term of the life of the Peacock is twenty-five years, according to the ancients †; and this determination feems to be well-founded, fince the bird is full grown before the end of three years, and the feathered race attain to a greater age than quadrupeds, becaufe their bones are more pliant. But I am furprifed that Willughby imagines, on the authority of Ælian, that the Peacock lived a complete century, efpecially as the account of that relator is mingled with many circumftances evidently fabulous ‡.

I have already faid, that the Peacock feeds on all forts of grain, like the gallinaceous tribe. The ancients generally gave it a monthly allowance of a bufhel of wheat, weighing about twenty pounds. It is proper to notice that the flower of the elder is hurtful to them §, and that the leaf of the nettle is, according to Franzius, a mortal poifon to the young Peacocks.

As in India the Peacocks live in the ftate of nature, it is ufual in that country to hunt them. They can hardly be approached in the day-time, though they are fcattered over the

* Belon.   † Ariftotle, Hift. Anim. vi. 9.—Pliny, x. 20.
‡ Ælian de Nat. Anim. xi. 33.   § Linnæus.

fields

fields in numerous flocks; becaufe, as foon as
they defcry a fportfman, they fly away more
fpeedily than partridges, and conceal themfelves
in the thickets, where they cannot be purfued.
The night therefore is the only proper time for
the chafe, which, in the vicinity of Cambaya
is conducted in the following manner:

The fportfmen get clofe to the tree where
the Peacocks are perched, and prefent a kind of
banner, which fupports two burning candles,
and is painted with the figures of Peacocks. The
Peacock dazzled by the glare, or engaged in admiring the painting, ftretches out its neck repeatedly, and again draws it back, and when its
head is obferved to be entangled in a running
knot, placed for the purpofe, the hunters
immediately draw the cord and fecure the
bird *.

We have feen that the Greeks much admired
the Peacock, but this was only for the beauty of
the plumage. The Romans, who carried every
luxury to excefs, actually feafted on Peacocks flefh. The orator Hortenfius was the
firft who ordered it to be ferved up at his table †,
and his example being followed, this bird came
to be fold at a very high price at Rome. The
Emperors refined on the luxury of their fubjects; and Vitellius and Heliogabalus gloried
in filling enormous chargers ‡ with the brains of

* Tavernier.   † Varro, *De Re Ruftica*, lib. iii. 6.
‡ Among others that called by Vitellius the Ægis of Minerva.

Peacocks,

Peacocks, the tongues of the *phænicopterus*, and the livers of the *fcarus* *, forming infipid difhes, whofe whole merit confifted in their deftructive expence.—In thofe times, a flock of an hundred Peacocks could bring a revenue of 60,000 fefterces, three Peacocks being only required of the keeper for each hatch †. This fum, according to the eftimation of Gaffendi, amounts to 10 or 12,000 livres. Among the Greeks, the cock and hen together coft a thoufand *drachmæ*, which correfponds to eighty-feven livres ten fous on the higheft valuation, twenty-four livres on the loweft. But the laft was undoubtedly reckoned much under value; elfe the exclamation in Athenæus would have no meaning:—" Is it not madnefs to rear Peacocks, " when they are as dear as ftatues ‡ ?'' The price muft have greatly fallen towards the beginning of the fixteenth century; fince in the " Nouvelle Coutume de Bourbonnois," publifhed in 1521, the Peacock is valued at two fous fix deniers money of that time, which Dupré de Saint Maur values at three livres fifteen fous of the prefent currency. But it would feem that foon after this period the price was advanced; for Bruyere tells us, that in the neighbourhood of Lifieux, where they could eafily rear Peacocks with the cyder lees, they bred flocks, which were very profitable, fince, being rare in other

---

* Suetonius.   † Varro, *De Re Rufticâ*, lib. iii. 6.

‡ Anaxandrides apud Athenæum, lib. xiv. 25.

parts of the kingdom, they were ufually fent from thence to all the confiderable cities, to be ferved up in fplendid entertainments. However, fcarce any but young ones are fit to be eaten; for their flefh is naturally dry, and grows hard as they become old. To this quality we muft impute the fingular property, which appears well afcertained, that their flefh can be kept feveral years without putrifying *. Yet old ones have been ufed, though more for fhow than ufe ; for they were ferved up decorated with their richeft plumes †. This is a well imagined refinement in luxury, and which the induftrious elegance of the moderns has added to the extravagant magnificence of the ancients. It was over a Peacock dreffed in this way, that our old knights made, on grand occafions, the vow called the *Vow of the Peacock* ‡.

Peacock's feathers were formerly ufed to make a fort of fans §, and they were formed into crowns like thofe of laurel, for the *Troubadour* poets. Gefner ‖ faw a web whofe woof was filk and gold thread, and the warp Peacocks feathers. Such no doubt was the robe woven with thefe feathers which Pope Paul III. fent to king Pepin ¶.

* St. Auguftine, *de Civitate Dei*, lib. xxi. 4.   † Aldrovandus.
‡ Mem. de l'Acad. des Infcript. tome xx. 636.   § Frifch.
‖ *Traite de Tournois*, par le pere Meneftrier.
¶ Genealogie de Montmorency.

According

According to Aldrovandus, Peacocks' eggs are reckoned by the moderns as improper food; whereas the ancients put them in the firſt claſs, and even before thoſe of the gooſe and common hen *. This contradiction he explains by ſaying, that they are pleaſant to the taſte, but pernicious to the health. It remains to be inquired whether the temperature of the climate affects their quality. [A]

* Athenæus.

[A] Specific character of the Peacock, *Pavo Criſtatus* :—" Has " a compreſſed creſt on its head; with ſingle ſpurs.''

The WHITE PEACOCK.

Climate has no leſs influence on the plumage of birds than on the fur of animals. We have elſewhere ſeen, that the hare, the ermine, and moſt other animals are ſubject to grow white in cold countries, particularly in the winter ſeaſon. Here is a ſpecies or a variety of Peacocks, which ſeems to have received ſimilar impreſſions from the ſame cauſe: and the effects are even greater, ſince the race is permanent; for the whiteneſs of hares and ermines is merely temporary, and happens only in the winter, like that of the ptarmigan. The colour of the White

9

Peacock, on the other hand, is no longer affected by the feafon or climate, and the eggs hatched even in Italy produced a white brood. The one which Aldrovandus has caufed to be engraved, was reared at Bologna; and this circumftance has made him fufpect that this variety did not belong peculiarly to cold countries. Yet moft naturalifts agree in affigning Norway and other northern countries for its native region *. It would feem that it is there wild, for in the winter it travels into Germany, where it is commonly caught in that feafon †. They are indeed found in countries much farther fouth, as in France and Italy, but there they are in the domeftic ftate ‡.

Linnæus affirms in general, as I have before faid, that Peacocks are averfe to refide in Sweden, and he excepts not even the white fort.

It required a long period of time, and a fingular concurrence of circumftances, to reconcile a bird, bred in the delicious climates of Afia and India, to the rigours of the northern tracts. If it had not been carried thither, it could not have migrated to thefe inhofpitable countries, either by the north of Afia, or by the north of Europe.

Though the date of this event be not exactly known, I prefume that it is not very diftant;

---

* Frifch, and Willughby.   † Frifch.

‡ Aldrovandus. He adds alfo the Madeira iflands, citing *Cadamofto de Navigatione*.

for,

for, on the one hand, I learn from Aldrovandus, Longolius, Scaliger, and Schwenckfeld, that it is not long since White Peacocks were esteemed as rarities; and on the other hand, I have grounds to believe that the Greeks were unacquainted with them, because Aristotle, having spoken in his *Treatise on the Generation of Animals* of the variegated colours of the Peacock, and afterwards of white partridges, white ravens, and white sparrows, takes no notice of White Peacocks.

The moderns add nothing to the history of this sort of Peacocks, except that the young are very delicate and difficult to rear *. It is however likely, that the influence of climate is not confined to the change of plumage alone, but must have operated in some degree on their temperament, instincts, and habits. I am surprised that no naturalist has observed the progress of the alterations, or at least noticed the intimate and latent effects produced. A single discovery of this kind would undoubtedly be more interesting, and tend more to the improvement and extension of natural knowledge, than the minute enumeration of all the feathers of these birds, and the laborious description of all their shades and tints, in the four quarters of the world.

Lastly, though their plumage be entirely white, and particularly the long feathers of the

* Schwenckfeld.

tail,

tail, we can ftill perceive at their extremities diftinct traces of thofe fpangles which formed their fineft ornament, fo deep was the impreffion of their primæval colours * ! It would be a curious fubject to try to revive thefe colours, and to determine by experiment what length of time, and how many generations would be required, in a fuitable climate, fuch as that of India, to reftore them to their original luftre.

* Frifch.

The VARIEGATED PEACOCK.

Frifch fuppofes that this is produced by the union of the common Peacock with the white kind. It bears indeed on its plumage the impreffion of this origin ; for white is fpread on its belly, its wings, and its cheeks. In the reft of the body, it is like the common Peacock, except in the fpangles of the tail, which are neither fo broad, fo round, nor fo well defined. All that I can find in authors with refpect to the particular hiftory of this bird is, that the young ones are not fo delicate in rearing, as thofe of the White Peacock,

The COMMON PHEASANT *.

Le Faifan, Buff.
Phafianus Colchicus, Linn. Gmel. &c.
Phafianus, Briff. Frifch, Gefner, &c

THE name of this bird is alone fufficient to indicate its native country. The Pheafant, or the Bird of Phafis, was confined, it is faid, to Colchis, before the expedition of the Argonauts †. That bold body of adventurers faw, in afcending the Phafis, thefe beautiful birds fcattered along its banks; they carried them home to Greece, and in doing fo they conferred a richer prefent than that of the golden fleece.

Even at prefent the Pheafants of Colchis or Mingrelia, and fome other countries bordering on the Cafpian, are the fineft and largeft that are known ‡. From thence they have fpread weftward through Greece, from the fhores of

---

* In Greek, Φασιανος; in Latin alfo, *Phafianus*; in Turkifh, *Surghun*; in Italian, *Fafiano*; in German, *Fofan*.

† " Argivâ primum fum tranfportata carinâ,
    " Ante mihi notum nil, nifi Phafis, erat." MARTIAL.

‡ Marco Polo affirms, that the countries fubject to the Tartars breed the largeft Pheafants, and thofe which have the longeft tail.

the

THE COMMON PHEASANT.

the Baltic * to the Cape of Good Hope †, and the ifland of Madagafcar ‡; and eaftward, through Media, to the remoteft parts of China § and Japan ‖, and even into Tartary. I fay through Media, for it appears that that country, which is congenial to the nature of birds, and which is ftocked with the moft excellent poultry, the moft beautiful peacocks, &c. has alfo proved a nurfery of Pheafants, and has fupplied many other regions ¶. They are exceedingly numerous in Africa, efpecially on the Slave Coaft **, the Gold Coaft ††, the Ivory Coaft, the country of Iffini ‡‡; the kingdoms of Congo and Angola §§, where the Negroes call them *Galignoles*. They are pretty common in different parts of Europe; in Spain, Italy, efpecially in the Pope's dominions, the

---

* Regnard killed one in the forefts of Bothnia. See his Voyage to Lapland.

† We perceive no difference between the Pheafants of the Cape of Good Hope and ours. KOLBEN.

‡ *Defcription de Madagafcar par Rennefort.* There is in Madagafcar a number of large Pheafants. FLACCOURT, *Hiftoire de Madagafcar.*

§ *Voyages de Gerbillon.* In the Corea we fee abundance of pheafants, hens, larks, &c. HAMEL, *Relation de la Corée.*

‖ Kœmpfer fays that at Japan there are Pheafants of great beauty.

¶ Athenæus relates, that thefe birds were fent for from Media, as being more numerous and of a better kind. ALDROVANDUS.

** Bofman's Defcription of Guinea.

†† Villault de Bellefond, *Relation des côtes d'Afrique.*

‡‡ Loyer in the *Hift. Gen. des Voyages.*

§§ Pigafetta.

Milanefe,

Milanefe *, fome iflands in the Gulph of
Naples, in Germany, France, England †; but
in the two laft countries they are not generally
met with. The Authors of the Britifh Zoology
affure us that, in the whole extent of Great
Britain, there is not a fingle Wild Pheafant.
Sibbald agrees with thefe naturalifts, fince he
tells us that in Scotland fome gentlemen breed
thefe birds in their houfes ‡. Boter affirms ftill
more directly, that there are no Pheafants in
Ireland § Linnæus takes no notice of them
in the enumeration he has given of the Swedifh
birds. In the time of Schwenckfeld, they were
very rare in Silefia; and it is only twenty years
fince they were introduced into Pruffia ||, though
they are very frequent in Bohemia ¶. If they
have multiplied in Saxony, it is owing to the
attention of the Duke Frederic **, who let
loofe two hundred in that country, and pro-
hibited their being caught or killed. Gefner,
who travelled through the mountains of Switzerland, affirms that he never faw any. It is true,
indeed, that Stumpfius afferts the contrary; but
it is probable that they may be found in fome
diftricts which Gefner had not examined, as in
that part which borders on the Milanefe territories, where Olina fays they are very common.

* Olina and Aldrovandus.   † Hiftory of Harwich.
‡ *Prodromus Hiftoriæ Naturalis Scotiæ.*
§ Willughby. || Klein.   ¶ Id.
** Aldrovandus.

The

Pheafants are far from being plentiful in
France. In the northern provinces they are
feldom feen, and would perhaps foon die away
but for the attention beftowed on the prefervation
of the royal game. Even in Brie, where fome
are continually making their efcape from their
keepers, and where their nefts, with eggs, have
been found in the extenfive forefts of that province, fo unfavourable is the climate that the
number of the wild Pheafants is never obferved
to increafe. We knew an opulent perfon
in Burgundy, who was at the utmoft pains
and fpared no expence in ftocking his eftate,
which lay in Auxois, but without fuccefs. I
fhould therefore fufpect that Regnard muft
have been miftaken when he tells us, that he
killed two Pheafants in Bothnia *; and Olaus
Magnus, who fays that they are found in Scandinavia, where they lie under the fnow through
the winter without any fuftenance †. This
habit feems to belong rather to the grous than
to the Pheafants; and the name *Gallæ fylveftres*,
which Olaus applies, fuits better that genus of
birds. My conjecture has the more foundation,
fince neither Linnæus, nor any other accurate
obferver, mentions feeing real Pheafants in the
northern countries. In fhort, we may fuppofe,
that the name Pheafant has firft been given by

---

* Regnard, *Voyage de Lapponie*.
† Quoted by Aldrovandus.

the natives to the grous, which are very numerous in the boreal tracts, and afterwards adopted blindly by travellers, and even by compilers, who are equally inattentive in difcriminating fpecies.

Since the wings of the Pheafant are fhort, confequently its flight low and laborious, we may readily conclude, that it could not traverfe the immenfe ocean that divides America from the temperate countries in the Ancient Continent. Accordingly none have been found in the New World, but only fome birds a-kin to them. I fpeak not of the true Pheafants which are at prefent common in the plantations of St. Domingo; for thefe, as well as the peacocks and pintados, were introduced by the Europeans *.

The Pheafant is of the fize of the common cock †, and in fome refpects rivals the peacock in beauty. His figure is as dignified, his deportment as bold, and his plumage almoft as refplendent. The colours of the Chinefe Pheafant are even brighter; but he has not, like the peacock, the power of difplaying his rich plumage, and of elevating the long feathers of his tail. Befides, the Pheafant has neither the creft of the peacock, nor the double tail; of

---

* *Hiftoire de l'Ifle Efpagnole de St. Domingue.*

† Aldrovandus, who has carefully obferved and defcribed this bird, fays, that he examined one which weighed three pounds of twelve ounces, *libras tres duodecim unciarum*, which fome have ignorantly tranflated, three pounds twelve ounces.

which

which the shorter one consists of quills capable
of being erected, and the longer one formed
of the coverts of these: in general, the Pheasant seems to have been modelled after less
slender and less elegant proportions; the body
thicker, the neck shorter, the head larger,
&c.

The most remarkable traits in its appearance
are, the two spots of scarlet in the middle of
which the eyes are placed, and the two tufts of
feathers of a gold-green, which, in the love
season, rise on each side under the ears; for in
animals there is almost always, as I have already
remarked, a new production, more or less remarkable, which is a sign that the generative
faculty is again roused to action. These tufts
of feathers are probably what Pliny calls sometimes ears *, sometimes little horns †. A prominence is observed at their base, formed by an
erector muscle ‡. Besides these, the Pheasant
is furnished with feathers at each ear, to close at
pleasure the orifice, which is very large §.

The feathers of the tail and rump have their
ends heart-shaped, like some of the tail-feathers
of the peacock ‖.

I shall not here enter into a particular description of the colours of the plumage; I shall only

---

* *Geminas ex plumâ aures submittunt subriguntque.* Lib. x. 48.
† *Phasianæ corniculis.* Lib. xi. 37.   ‡ Aldrovandus.
§ Ibid.   ‖ Brisson.

observe,

obferve, that in the female they are much lefs brilliant than in the male, in whom the reflexions are ftill more fugitive than in the peacock, and depend not only on the various incidence of the light, but on the junction and pofition of the feathers : for if any one be taken fingly, the green wavings vanifh, and we fee only a brown or black *. The fhafts of the feathers of the neck and the back are of a fine bright yellow, and appear like fo many plates of gold †. The coverts under the tail continue diminifhing, and terminate in a kind of filaments. The tail confifts of eighteen quills, though Schwenckfeld reckons only fixteen ; the two middle ones are the longeft of all, and they fhorten regularly towards the fides. Each leg is furnifhed with a fhort pointed fpur, which has efcaped fome defigners, and even the engraver of our *Planches Enluminées*, No. 121 ; the toes are connected by a membrane broader than ufual in pulverulent birds ‡, and feems to form the firft fhade between thefe and the aquatic tribes ; and in fact Aldrovandus obferves, that the Pheafants delight in wet places ; and he adds, that they are fometimes caught in the marfhes in the neighbourhood of Bologna. Olina, another Italian, and Le Roi, Lieutenant of Rangers at Verfailles, have made the fame remark. The laft-mentioned perfon informs me, that it is always in

* Aldrovandus.   † Ibid.   ‡ Ibid.

the

the moft watery fpots, and along the fides of
the pools in the large forefts of Brie, that Pheafants lodge which have efcaped from the hunters
in the vicinity. Though habituated to the fociety of man, though loaded with his favours,
thefe Pheafants retire as far as poffible from all
human dwellings; for thefe birds are very
wild, and extremely difficult to tame. It is
faid, however, that they can be inftructed to
return at the found of a whiftle *; that is, they
can be attracted by this means to their food;
but as foon as their appetite is fatisfied, they
return to their natural mode of life, and forget the hand that fed them. They are ftub-
born flaves, that will not fubmit to conftraint,
who know nothing defirable that can enter into
competition with liberty; who feek continually
to recover it, and never lofe fight of it when
opportunity occurs †. The wild ones newly
bereaved of freedom become furious; they dart
with violence on the companions of their captivity, and ftrike with their bills, nor do they
fpare even the peacocks ‡.

---

* *Journal Economique mois de September* 1753. It is very probable that this was all the attainment of the tame Pheafants,
which, according to Ælian, were bred in the *menagerie* of the
King of India, lib. xviii.

† " Though reared in the houfe, and hatched under a hen,
" they never grow domeftic, but ftill retain their rufticity."
OLINA.——Which confirms what I have myfelf obferved.

‡ Longolius, *apud Aldrovandum.*

Thefe

These birds are fond of living in woods that grow on the plains, differing in this respect from the grous, which inhabit forests that clothe the mountains. They perch on the tops of trees during the night *, sleeping with their head under the wing; their cry, that is the cry of the male, (for the female has none at all,) is intermediate between that of the peacock and the pintado, but more like that of the latter, and therefore far from being agreeable.

Their disposition is so unsocial, that they not only fly from the presence of man, but avoid the company of each other, except in the months of March and April, when the male courts the female. It is then easy to discover them in the woods, because they are betrayed by the loud noise made by the clapping of their wings, which may be heard at a great distance †. The Cock Pheasants are not so ardent as the common cocks. Frisch asserts that, in the wild state, each attaches itself to a single female: but man, who glories in perverting the order of nature to his interest or his whims, has changed the instinct of these birds, by habituating each cock to serve seven hens, and constraining these to rest satisfied with the performance of a single male.

Some have had patience to make all the observations necessary to determine this pro-

* Frisch.   † Olina.

portion

portion to be the moſt profitable for breeding *.
Several œconomiſts, however, allow only two
females to each male †; ·and I muſt confeſs that
this diſtinction ſucceeded the beſt in ſome trials
I have made. But the different combinations
muſt depend on particular circumſtances; on the
temperature of the climate, the nature of the
ſoil, the quality and quantity of the food, the extent and poſition of the place for rearing them,
and the attention of the keeper, who ought to
remove the hen as ſoon as ſhe has imbibed the
quickening influence, and preſent the females
one after another at proper intervals. He
ſhould alſo give the cock during that ſeaſon
buck-wheat and other ſtimulating aliments, as is
uſual about the end of winter, when we want to
anticipate the period of love.

The Hen Pheaſant conſtructs her neſt alone;
ſhe ſelects the darkeſt corner of her lodging, and
forms it with ſtraw, leaves, and other materials;
though it appears very rude and unſhapely, ſhe
prefers it to any other not built by herſelf; inſomuch that if one be prepared for her of a regular
conſtruction, ſhe tears it in pieces, and arranges
the materials anew in her own way. She breeds
but once a year, at leaſt in our climates; ſhe lays
twenty eggs ‡ according to ſome, and forty or
fifty according to others, eſpecially if we ſave

---

* Journal Economique, Sept. 1753. Alſo ſee the article
Faiſanderie in the Encyclopedie.

† Friſch.—Maiſon Ruſtique.   ‡ Palladius, *De Re Ruſtica*.

her

her the trouble of fitting *. Thofe, however, which I had occafion to fee, never laid more than twelve eggs, and fometimes lefs, though thefe were hatched by common hens. They generally lay one every two or three days, and the eggs are much fmaller than thofe of an ordinary hen, and the fhell thinner even than thofe of pigeons. The colour is a greenifh-grey, fpeckled with little brown fpots, as Ariftotle has well obferved †, ranged in a circular zone round the egg. A Hen Pheafant can hatch eighteen.

If we would undertake to raife Pheafants on a great fcale, we muft for that purpofe allot a park of proportional extent, which fhould be partly laid out in grafs, and partly planted with bufhes, where thefe birds may be fhaded from the fun, fheltered from rain, and even protected from the affaults of the ravenous tribes. One part of this park ought to be divided into feveral fmall patches of ten or twelve yards fquare, conftructed fo that each may lodge a cock with his females, and they muft be confined either by difabling their wings, or by fpreading a net over the little inclofure. Care fhould be taken not to fhut up feveral cocks together; for they will undoubtedly fight, and perhaps kill each other ‡. We muft even contrive that they fhall not fee

* Journal Economique, Sept. 1753.
† Hift. Anim. lib. vi. 2. Imitated by Pliny, lib. x. 52.
‡ Journal Economique, Sept. 1753.

or

or hear each other, for though naturally cold
and phlegmatic, their difquietude or jealoufy
will interrupt or relax their amours. Thus, in
fome animals, as well as in man, jealoufy is not
always proportioned to the appetite of love.

Palladius alleges, that the cocks need only be
a year old *, and all naturalifts agree, that hens
are proper for breeding the third year. Sometimes when Pheafants are numerous, it is fufficient to lodge the females in the inclofures, and
leave them to the embraces of the wild cocks.

Thefe birds feed on all forts of grain and
herbs. It is even recommended to throw part of
the park into a kitchen garden, in which to raife
beans, carrots, potatoes, onions, lettuces, par-
fnips, and efpecially the two laft, of which they
are remarkably fond. It is alfo faid that they
love acorns, the berries of the white thorn, and
feed of wormwood †; but the food beft adapted
to them is wheat mixed with ants eggs. Some
advife not to mix ants themfelves, left they take
a diflike to the eggs; but Edmond King recommends the ants themfelves, and affirms that thefe
infects afford them the moft falutary nourifhment,
and can even reftore them when they are fickly
and drooping; and that inftead of thefe, we
may fubftitute even grafhoppers, ear-wigs, and
millepedes. The Englifh author, whom I have
juft quoted, affures us, that he loft many Phea-

* Journal Economique, Sept. 1753.
† Gerbillon, *Voyage de la Chine & de la Tartarie.*

fants

fants before he learnt this fact, but that after he attended to that circumftance, not one died of thofe which he was breeding *. But whatever fort of food we give them, it muft be offered fparingly, not to make them too fat; for corpulence blunts the ardor of the cock, weakens the prolific powers of the hen, and makes her lay eggs with foft fhells and eafily broken.

The time of incubation is from twenty to twenty-five days, according to moft authors and my own obfervation †. Palladius fixes it at thirty; but this is a miftake which ought not to have been adopted in the *Maifon Ruftique;* for in the warm climate of Italy, the Pheafants could not require fo long time to hatch, and therefore inftead of *trigefimus*, we ought to read *vigefimus.*

We ought to keep the fitting-hen in a place remote from noife and fomewhat under ground, fo as not to be affected by the variations of the weather, or expofed to the ftroke of thunder.

As foon as the young Pheafants leave the fhell, they begin to run like all the gallinaceous tribe. For the firft twenty-four hours, food is generally withheld from them; after that, they are put with the mother into a crib, and carried out every day to the fields, into the pafture grounds where ant-hills abounds. This ought

---

* Philofophical Tranfactions, No. 23.
† Gefner, Schwenckfeld.—Journal Economique, & le Roi.

to be covered with deals, which may be removed or replaced as occafion requires. It
ought alfo to have a divifion near one of the
ends, where the mother fhould be confined with
bars fo wide afunder however as to allow the
chickens to go out and return as often as they
chufe. The clucking of the imprifoned mother,
and the neceffity of being frequently warmed,
will conftantly bring them back and prevent them
from fauntering too far. It is ufual to join together three or four hatches of nearly the fame
age, fo as to form a fingle family, which may
be reared by the fame mother.

They are fed at firft, like all young chicks,
with a mixture of hard eggs, crumbs of bread,
and lettuce leaves mixed together, and with an
addition of the eggs of meadow ants. But at
this tender age two precautions muft be carefully
obferved. They muft not be allowed to drink at
all, nor be carried abroad till the dew is entirely
gone, for humidity of every kind is hurtful to
them. We may notice by the way, that this is
one of the reafons why hatches of wild Pheafants
feldom fucceed in France; for, as I have already
remarked, thefe birds prefer the frefh verdant
places, and in fuch fituations the young can
hardly furvive the damps. The fecond point to
be attended to is, that their food fhould be given
frequently and in fmall quantities, beginning as
foon as day break, and always mixing with it
ants eggs.

In

In the second month, more subſtantial
nouriſhment may be given; eggs of the wood
ants, turkey beans, wheat, barley, millet, ground
beans; and the intervals between the meals
may be gradually enlarged.

At this time they begin to be ſubject to ver-
min. To prevent that diſorder, moſt modern
writers adviſe us to clean the crib, or even to
lay it aſide altogether, except the ſmall roof
which ſerves to ſhelter them. Olina recommends a plan propoſed by Ariſtotle, which ſeems
to me better contrived and more ſuitable to the
nature of theſe birds. They are in the number
of thoſe that welter in the duſt, and when that
gratification is withheld, they languiſh and die *.
Olina directs ſmall heaps of dry earth or very fine
ſand to be laid near them, in which they may
tumble and rid themſelves of the painful itching
occaſioned by the infects.

We muſt alſo be very attentive in giving them
clean water, and in often renewing it, elſe they
will be in danger of contracting the pip, of which
there is ſcarcely any remedy, according to the
moderns; though Palladius adviſes to remove it
as in common chickens, and to rub the bill with
garlick bruiſed with tar.

The third month is attended with new diſeaſes.
The tail feathers then drop and others appear,
which is a ſort of criſis to them, as well as to

* ARIST. Hiſt. Anim. lib. v. 31.

the

the Peacocks. But ants eggs are ftill a refource ;
they haften the trying moment, and leffen the
danger, provided we do not give them too much,
for the excefs is pernicious. In proportion as
the young Pheafants grow up, their regimen
becomes the more like that of the adults; and
at the end of the third month, they may be let
loofe in the place intended to be ftocked. But
fuch is the effect of domeftication on animals that
have lived fome time in that ftate, that even thofe
which, like the Pheafants, have an invincible
attachment to liberty, cannot be reftored to it
but by imperceptible degrees; in the fame
manner as a good ftomach that has been weak-
ened with watery elements, cannot at once recover its tone, fo as to digeft rich food. We
muft firft carry the crib which contains the brood
to the field where the colony is to be difperfed ;
we muft give them what food they like beft, but
never in the fame fpot; and we muft diminifh
the quantity every day, and thus by degrees
conftrain them to provide for themfelves, and to
become acquainted with the country. When
they are able to procure fubfiftence, they fhould
be refigned to liberty and nature. They will
foon grow as wild as thofe bred in the woods ;
except only that they will ftill retain a fort of
affection for thofe fpots where they were foftered in their infancy.

Man, encouraged by his fuccefs in changing
the inftinct of the Cock Pheafant, and in recon-
ciling

ciling it to the fociety of a number of females, has tried alfo to effect another violence, to make it breed with a foreign fpecies; and the experiments have in fome degree fucceeded, though they required great care and attention *. A young Cock Pheafant which had never copulated, was fhut in a clofe place where but a faint light glimmered through the roof: fome young pullets were felected, whofe plumage refembled the moft that of the Pheafant, and were put in a crib adjoining that of the Cock Pheafant, and feparated from it only by a grate, of which the ribs were fo clofe as to admit no more than the head and neck of thefe birds. The Cock Pheafant was thus accuftomed to fee thefe females, and even to live with them, becaufe the food was thrown into the crib only. When they had grown familiar and the feafon of love approached, both the cock and hens were fed on heating aliments, to provoke their defires; and after they difcovered an inclination to couple, the grate which parts them was removed. It fometimes happened, that the Cock Pheafant, faithful to nature and indignant at the infult offered him, abufed the hens, and even killed the firft he met with: but if his rage did not fubfide, he was on the one hand mollified by touching his bill

* The Wild Pheafants never tread the hens which they meet; not but they fometimes make advances, only the hens will never permit them to proceed. I owe this, among many other obfervations, to M. Le Roi, Lieutenant des Chaffes at Verfailles.

with

with a red-hot iron, and on the other, stimulated by the application of proper fomentations. At last his appetites however growing every day more fiery, and nature constantly counteracting herself, he at last copulated with the hens, which in consequence laid eggs dotted with black, like those of the Pheasant, but much larger; and they produced hybrids partaking the properties of both species, and, according to some, more delicate, and even better flavoured than the true sort, but incapable, it is said, of propagating their kind: yet Longolius asserts, that the females of this kind which couple with their sire, produce real Pheasants. Care has also been taken to give the Cock Pheasant only virgin hens; whether the more to incite the males, (for man judges of all creatures from himself,) or because the repetition of the experiment on the same subjects is said to occasion the breed to degenerate.

It is pretended that the Pheasant is a stupid bird, and imagines itself safe when its head is concealed; which has been alleged of many other birds that heedlessly fall into all sorts of snares. When hunted by a pointer, and met, it stands still, and looks steadily at the dog, so that the sportsman can take his aim at leisure. To decoy it, we need only present its own figure, or a red rag on a white sheet. It is caught also by setting gins in the tracks which it treads in the morning to drink. It is also chased by the

7   falcon,

falcon, and such as are taken this way, are said to be more delicate and delicious than ordinary*. Autumn is the season when they are fattest. The young ones may be fattened like other poultry, only in introducing the little ball into the throat, care should be taken to prevent the tongue from being pushed backwards, which would infallibly kill the bird.

A fat young Pheasant is a most exquisite morsel, and at the same time very wholesome food. Accordingly this luxury has been always reserved for the tables of the rich, and the whim of Heliogabalus of feeding his lions on Pheasants, has been regarded as the most wanton profusion.

According to Olina and Le Roi, this bird, like the common hens, lives about six or seven years; but the opinion that the age may be discovered from the number of the cross bars on its tail, is void of foundation.   [A]

* Aldrovandus.

[A] Specific character of the Pheasant, *Phasianus Colchicus*: " It is rufous, its head blue, its tail wedge-shaped, its cheeks " marked with *papillæ*."

The WHITE PHEASANT.

Phasianus Colchicus Albus, Linn.

We are not sufficiently acquainted with the history of this variety, to determine the cause to which we ought to refer the whitenefs of its plumage: analogy would lead us to fuppofe it to be the effect of cold, as in the cafe of the White Peacock. It is true, that the Pheafant has not been introduced fo far into the northern regions as the Peacock; but the white is alfo not fo pure, fince, according to Briffon, it has fpots of deep violet on its neck, and other rufty fpots on the back; and according to Olina, the males have fometimes the full colours of ordinary Pheafants on the head and neck. This laft author afferts, that the White Pheafants come from Flanders; but in Flanders they undoubtedly fay, that they come ftill farther north. He fubjoins that the females are of a purer white than the males; and I have myfelf obferved that property to obtain in the Pheafants.

The VARIEGATED PHEASANT.

Phafianus Colchicus Varius, Linn.

As the White Peacock, when coupled with the common fort, produces the variegated kind, we may fuppofe that the White and the common Pheafant would breed the variety here mentioned; efpecially as it has the fhape and even the fize of the ordinary fort, and its plumage, the ground of which is white, is fprinkled with fpots that have all the ufual colours.

Frifch

Frifch obferves, that the variegated Pheafant is not proper for propagation.

The COCQUAR, or BASTARD PHEASANT.

Phafianus Colchicus Hybridus, Linn.
The Hybridal Pheafant, Lath.
The Pied Pheafant, Hayes.

The name which Frifch gives to this variety fhews that he confidered it as bred between the Cock Pheafant and the common hen.   It refembles indeed the Pheafant, by the red circle round its eyes, and its long tail; and it approaches the common cock, by the dull and homely feathers of its plumage.   It is alfo fmaller than the ordinary Pheafant, and like the other Hybrids it is incapable of producing its fpecies.

Frifch tells us, that many of thefe are raifed in Germany, being profitable ; and that they are excellent food.

THE PIED PHEASANT.

FOREIGN BIRDS
ANALOGOUS TO THE PHEASANT.

I SHALL not range under this denomination, feveral birds on which moft travellers or naturalifts have beftowed the name of Pheafant, but which, after a clofe inveftigation, we have determined to belong to very different tribes.— Such as, 1. The Pheafant of the Antilles of Briffon, which is that of the ifland Kayriouacou of Father Tertre, and which has longer legs and a fhorter tail than the Pheafant. 2. Briffon's crowned Pheafant of the Indies, which differs from the Pheafant by its general form, and by the fhape of its bill, its inftincts and habits, its long wings and fhort tail, and which, if we except its fize, feems to refemble much the pigeons. 3. The American bird, which we have directed to be figured under the name of *The Crefted Pheafant of Cayenne*, becaufe it was fent to us under that name; but which appears to be diftinguifhed from the Pheafant by its bulk, its carriage, its long flender neck, its fmall head, its long wings, &c. 4. The *Hocco Pheafant* of Guiana, which is by no means a Pheafant, as the comparifon of the figures alone fuffices to fhew. 5. All the other *Hoccos* of Ame-

rica,

rica, which Briffon and Barrere, and others who
have been mifled by their fyftems, have referred
to the genus of the Pheafant ; though they differ
in many refpects, and even in fome properties
that have been received as generic characters.

I.

The PAINTED PHEASANT.

Faifan Doré, ou *Le Tricolor Huppé de la Chine*, Buff.
Phafianus Pictus, Linn. and Gmel.
Phafianus Sanguineus, Klein.
Phafianus Aureus Sinenfis, Briff.
Gold Fafian, Gunth.

Some authors, who have applied to this bird
the name of *Red Pheafant*, would have had equal
reafon to have called it the *Blue Pheafant*, and
the term *Golden Pheafant* is equally inadequate
to denote the plumage, which is enriched by the
luftre of all thefe three colours.

It may be confidered as a variety of the ordinary fpecies, whofe garb fparkles with the decorations of a happier clime. They are two
branches of the fame family, which, though long
feparated, recal their common defcent, and can
ftill intermingle, and breed with each other.
But it muft be confeffed-that their progeny par-
takes fomewhat of the fterility of Hybrids; which
proves

proves the antiquity of the partition of the paternal houfe.

The Painted Pheafant is fmaller than the ordinary Pheafant. The remarkable beauty of this bird has occafioned its being fo much bred in our pheafant walks. The predominant colours of its plumage are red, gold, yellow, and blue; it has long beautiful feathers on the head, which can be erected at pleafure; its iris, bill, legs, and nails, are yellow; the tail is proportionally longer than that of the common Pheafant, more mottled, and in general of a brighter plumage; above the feathers of the tail others are fpread long and narrow, and of a fcarlet colour, with a yellow fhaft; the eyes are not encircled with red fkin, like the European Pheafant: in a word it appears to have been deeply marked by the impreffion of the climate.

The female of the Painted Pheafant is fomewhat fmaller than the male, and its tail is not fo long; the colours of its plumage are very ordinary, and even inferior to thofe of the common kind; but fometimes they acquire in time the beauty of the male. In England, one belonging to Lady Effex changed, in the fpace of fix years, its mean dufky colour into the rich luftre of the male.; fo as not to be diftinguifhed, except by the appearance of the eyes and the length of the tail. Intelligent perfons who have had opportunities of obferving thefe birds, in-

form

form me, that this change of colour takes place in moft females, and begins at four years old, when males take a diflike to them and treat them harfhly. That then thofe long narrow feathers, which in the male lie over the tail, begin to appear. And in a word, as they grow older, they become the more like the males, which in a certain degree happens in all animals.

Edwards tells us, that he faw at the Duke of Leeds's, a common Hen Pheafant, whofe plumage had in the fame manner become like that of the male. He adds, that fuch changes of colours feldom take place except among birds that live in the domeftic ftate.

The eggs of the Painted Pheafant are very like thofe of the Pintado; they are proportionally fmaller than thofe of the domeftic Hen, and more reddifh than thofe of the common Pheafant.

Sir Hans Sloane kept a male about fifteen years: it would therefore feem that this bird is hardy, fince it lived fo long out of its native abode. It is foon reconciled to our climate, and multiplies faft; it breeds even with the European Pheafant. Le Roi, Lieutenant of the Rangers at Verfailles, put one of them to a Cock Pheafant of this country, and obtained two Cock Pheafants very like the common kind, but the plumage had a dirty caft, and only a few yellow feathers on the head like thofe of the Painted
Pheafant:

Pheafant: and thefe two young males being paired with European hen-pheafants, one fucceeded
the fecond year, and a hen-pheafant was hatched,
which could never be made to breed. The two
Cocks produced no more, and the fourth year
made their elopement.

It is probable that the Painted Pheafant is
that elegant pheafant whofe plumes fell higher
in China than the pheafant itfelf; and alfo the
fame with what *Marco Polo* admired in one of
his travels to China, whofe tail was two or three
feet long. [A]

[A] Specific chara&er of the Painted Pheafant, *Phafianus
Pictus :*—" Its creft is yellow, its breaft faffron, its fecondary
" wing-quills blue, its tail wedge-fhaped."

II.

The BLACK-AND-WHITE CHINA PHEASANT.

Phafianus Nycthemerus, Linn. and Gmel.
Phafianus Albus Sinenfis, Briff. and Klein.
Silber Fafian, Gunth.
The Pencilled Pheafant, Lath.

The figure in the *Planches Enluminées* was
taken from a ftuffed fpecimen; and I doubt not
but that of Edwards, which was drawn from
the life, and retouched at leifure, the minute

parts

parts being added from the dead fubject, reprefents this Pheafant more exactly, and gives a
better idea of its air and port, &c.

It is eafy to fee, from the bare infpection of
the figure, that it is a variety of the Pheafant,
having the general proportions of the Painted
Chinefe Pheafant, but larger, and exceeding
even the European kind. It refembles the laft
in a remarkable property, having a red border
round the eyes, which is even broader and of
greater extent; for it falls on each fide below
the under mandible, and at the fame time
rifes like a double comb above the upper mandible.

The female is rather fmaller than the male,
and differs much in colour. It has neither the
upper-fide of the body white, nor under-fide of
a fine black, with purple reflexions. In no part
of its plumage is there any white, except a
fingle fpeck below its eyes; the reft is of a
brown red, more or lefs deep, except under the
belly and on the lateral feathers of the tail,
where there are black tranfverfe bars on a gray
ground. In every other refpect there is lefs
difference between the fexes in this than in any
other Pheafant: the female has, like the male,
a tuft on its head, its eyes are encircled with
a red border, and its legs are of the fame
colour.

Since no naturalift, or traveller, has given
the leaft hint concerning the original abode of
the

the Black-and-white Pheafant, we are obliged
to form conjectures. I am inclined to fuppofe
that, as the Pheafant of Georgia, having migrated towards the eaft, and having fixed its refidence in the fouthern or temperate provinces
of China, has become the Painted Pheafant; fo
the White Pheafant, which is an inhabitant of
our cold climates, or that of Tartary, having
travelled into the northern provinces of China,
has become the pencilled kind : that it has there
grown to a greater fize than the original Pheafant, or that of Georgia ; becaufe it has found
in thefe provinces food more plentiful and better fuited to its nature : but that it betrays the
marks of a new climate in its air, port, and external form ; in all which it refembles the
Painted Pheafant ; but retains of the original
Pheafant the red orbits, which have been
even expanded from the fame caufes undoubtedly that promoted the growth of its body,
and gave it a fuperiority over the ordinary
Pheafant.   [A]

[A] Specific character of the Pencilled Pheafant, *Phafianus
Nycthemerus :—*" It is white, its creft and belly black, its tail
" wedge-fhaped."

III. The

III.

The A R G U S, or L U E N.

Phasianus Argus, Linn. and Gmel.
The Argus Pheasant, Lath.

In the north of China, another fort of Pheafant has been found, the wings and tail of which
are fprinkled with a multitude of round fpots
like eyes; whence it has received the name of
Argus. The two feathers in the middle of the
tail are very long, and projeƈt much beyond
the reft; it is of the fize of a turkey; its head
is covered with a double creft, which lies backwards *.

---

* In the Philofophical Tranfaƈtions, vol. LV. p. 88, for 1766,
is a very full defcription of this bird, accompanied with a good
engraving, framed by Mr. Edwards from a drawing fent from
China.

IV. The

IV.

The NAPAUL, or HORNED PHEASANT.

Meleagris Satyra, Linn.
Penelope Satyra, Gmel.
Phasianus Cornutus, Briff.
The Horned Pheasant, Lath.

Edwards, to whom we are indebted for our acquaintance with this uncommon bird, ranges it among the turkies, on account of the flefhy excrefcences on the head, and yet he has given it the name of Horned Pheafant. I fhould fuppofe that it is more like the pheafant than the turkey: for thefe protuberances are by no means peculiar to the turkey; they belong alfo to the cock, the pintado, the royal bird, the caffowary, and many others in both continents; nor are they even withheld from the pheafant, fince we may regard the broad circle of red fkin that furrounds the eyes, as nearly of the fame nature; and in the Pencilled Pheafant of China, this really forms the double comb on the bill, and the barbils under it. If we add, that the Napaul is an inhabitant of the congenial climate of pheafants, fince it was fent to Dr. Mead from Bengal; that in its bill, its feet, its fpurs, its wings, and its general form, it was like the pheafant; we fhall be convinced
that

that it is more natural to clafs it with the pheafants, than with an American bird fuch as the
turkey.

The Napaul, or Horned Pheafant, is fo called
becaufe of two protuberances which grow from
its head like horns, are of a blue colour, a cylin-
drical fhape, blunt at their ends, reclined backwards, and confift of a fubftance refembling
callous flefh. It has not that round circle about
its eyes which occurs in the pheafants, and is
fometimes dotted with black; the fpace which
furrounds the eyes, is fhaded with black hairs,
like feathers. Under this fpace, and from the
bottom of the lower mandible, grows a kind of
gorget confifting of loofe fkin, which falls
down and floats freely on the throat and the
upper part of the neck: this gorget is black
in the middle, and is fprinkled with a few
ftraggling hairs of the fame colour. It is
marked with wrinkles; fo that it appears to admit of extenfion in the living animal, and there
is reafon to fuppofe that it can be inflated or
contracted at pleafure. The lateral parts are
blue, with fome fpots of orange, and without
any hair on the outer furface; but the infide,
which applies to the neck, is fhaded with little
black feathers, as well as that part of the neck
which it covers. The crown of the head is red,
the fore-part of the body reddifh, and the hind-
part of a dufky colour. Over the whole bird,
including even the tail and the wings, we perceive

ceive white fpots, furrounded with black, and
difperfed with confiderable regularity: thefe
fpots are round on the fore-part, and oblong,
or fhaped like tears, on the hind-part, with the
point turned towards the head. The wings
fcarcely reach beyond the origin of the tail;
from which we may conclude that it is a heavy
bird. The length of the tail could not be
determined by Edwards, for in the original
drawing it is reprefented as being partly worn
off. [A]

[A] Specific chara&ter of the Horned Pheafant, *Penelope Satyra.*
—" It has a pair of horns on its head; its body is red, with fpang-
" ling points."

V.

The K A T R A C A.

Phafianus Motmot, Linn. and Gmel.
Phafianus Guianenfis, Briff.
The Motmot Pheafant, Lath.

Though there are no true pheafants in America, as we have already eftablifhed, yet among
the multitude of birds that inhabit that vaft
continent, fome poffefs the properties of that
tribe in a greater or lefs degree. The Katraca
approaches the neareft, and may be confidered

as

as the reprefentative of the pheafant in the
New World. Its general form, its bill, which
is flightly hooked, its eyes, which are encircled with red orbits, and its tail, which is remarkable for its length, are all characters which
prove it to be of a congenerous kind. At the
fame time, as it is a native of a diftant climate,
of even a different world, and as it is uncertain
whether it would breed with the European
pheafants, I range it in this place after the
Chinefe fort, which certainly couple with ours.
Its hiftory is totally unknown to us. We retain
the name *Katraca*, which, according to Father
Feuilleé, is the name it has in Mexico. [A]

[A] Specific character of the *Phafianus Motmot.*—" It is brown,
" below tawny, its tail wedge-fhaped, its lateral tail-quills
" rufous."

FOREIGN

[319]

FOREIGN BIRDS

THAT SEEM RELATED TO THE PEACOCK
AND PHEASANT.

☞ *I range under this vague title, some foreign birds, which have not been described with sufficient accuracy for us to assign their precise place.*

I.

The CHINQUIS.

Pavo Tibetanus, Linn. Gmel. and Briss.
The Thibet Peacock, Lath.

The name *Chinquis* is formed from the Chinese word *chin-tchien-khi.* The bird is the tenth species of the genus of Pheasant in Brisson's system. It is found in Thibet, whence that author has called it the *Thibet Peacock.* It is as large as the pintado; the iris is yellow, the bill ash-coloured, the feet gray, the ground of the plumage cinereous, variegated with black lines and white points. But its chief and distinguishing ornament is, the large round spots of brilliant blue, changing into violet and gold, spread, one by one, on the feathers of the back

4   and

and the coverts of the wings ; two and two, on the quills of the wings ; and four and four, on the long coverts of the tail, of which the two middle ones are the longeft; the lateral per-petually diminifh.

We are totally unacquainted with its hiftory; we are not even informed whether it expands its fine fpangled plumes into a fan.

We muft not confound the Chinquis with the Kinki, or Golden Hen of China, which is mentioned in the narrations of Navarette, Tri-gault, and du Halde ; and which, as far as we can judge from the imperfect accounts given of it, is nothing but the Painted Pheafant *.   [A]

* Abbe Prevot.   *Hift. Gen. des Voyages.*

[A] Specific character of the *Pavo Tibetanus :* — " It is ci-" nereous, ftriated with blackifh ; its head fomewhat crefted ; " with two fpurs."

II.

The S P I C I F E R E.

Pavo Muticus, Linn. and Gmel.
Pavo Japanenfis, Briff.
The Japan Peacock, Lath.

The Japan Peacock is the name given by Al-drovandus to what is referred to in the eighth fpecies of Pheafant by Briffon ; and both of thefe

thefe authors admit, that this bird refembles our peacock only by the feet and tail.

It has a fpike-fhaped tuft on its head ; for which reafon I term it *fpicifere*. This tuft is about four inches high, and appears enamelled with green and blue ; the bill is of an afh-colour, longer and more flender than that of the Peacock ; the iris is yellow, and the orbits red, as in the Pheafant ; the tail-feathers are fewer, their colour deeper, and their fpangles broader, but glowing with the fame tints as in the European Peacock. The diftribution of the colours forms on the breaft, the back, and that part of the wings next the back, a kind of fcales which give different reflexions in different places ; blue on the part of the wings next the back ; blue and green on the back ; blue, green, and gold-colour on the breaft : the other quills of the wing are green in the middle through their whole length, then yellowifh, and run into black at their extremity : the crown of the head, and the arch of the neck, are covered with blue fpots mixed with white on a greenifh ground.

Such is nearly the defcription which Aldrovandus has given of the male, from a painted figure fent by the Emperor of Japan to the Pope. He does not inform us whether it difplays its tail like our Peacock : but it is certainly

not fpread in Aldrovandus' figure; nor has it
any fpurs on the legs, though that author has
not omitted them in the engraving of the Common Peacock, which is placed oppofite to ferve
for comparifon.

According to Aldrovandus, the female is
fmaller than the male; has the fame colours on
the head, neck, breaft, and wings; but the
under-fide of its body is black, and the coverts
of the rump, which are much fhorter than the
quills of the tail, are decorated with four or five
fpangles of confiderable breadth in proportion
to the fize of the quills: green is the predominant colour in the tail, the feathers are edged
with blue, and their fhafts are white.—This
bird feems to be much akin to the bird which
Koempfer, in his Hiftory of Japan, mentions
under the name of Pheafant *. [A]

* " There is at Japan a kind of Pheafants diftinguifhed by the
" diverfity of their colours, by the brilliancy of their feathers,
" and by the beauty of their tail, which is as long as half a man's
" height, which, by this mixture and charming variety of the
" richeft colours, particularly of gold and azure, yields in no re-
" fpect to that of the Peacock." KOEMPFER.

[A] Specific character of the *Pavo Muticus:* " The creft
" on its head is awl-fhaped; no fpurs."

III. The

III.

The E P E R O N N I E R.

Pavo Bicalcaratus, Linn. and Gmel.
Pavo Sinenfis, Briff.
The Peacock Pheafant, Edw.
The Iris Pheafant, Lath.

This bird is hardly known, except from the figure and defcription which Edwards publifhed of the male and female, made from the living fubject.

At firft fight the male feems to bear fome analogy to the Pheafant and Peacock: like them it has a long tail, decorated with fpangles, as in the Peacock. And fome naturalifts, abiding by the firft impreffion, have ranged it with the Pheafants *. But though from the confideration of thefe exterior appearances, Edwards has been induced to retain the name of Peacock-Pheafant, he was convinced, on a clofer infpection, that it did not belong to the Pheafant kind: becaufe, 1. The long feathers of the tail are round, and not pointed at the end; 2. They are ftraight throughout, and not arched back; 3. They do not make an inverted gutter by the bending back of their webs as in the Phea-

* Klein and Briffon.

fant;

fant; 4. It does not walk with its tail raifed
and recurved as in that bird.

Still lefs does it belong to the Peacock kind,
from which it differs in the carriage of its tail,
in the difpofition and number of the quills that
compofe it. It is diftinguifhed too by other properties; its head and neck are thick, its tail does
not rife and fpread like the Peacock's *, and
inftead of a tuft, it has only a fort of flat creft
formed by the feathers on the top of the head,
which briftle and ftretch towards a point fomewhat projecting; and laftly, it has a double fpur
on each leg, a fingular character, from which
I have denominated the bird †.

Thefe external differences, which undoubtedly involve many others which are more concealed, would feem a fufficient reafon to every
fenfible man, who is not prejudiced by fyftems,
for excluding it from the Peacocks and Pheafants; though like thefe, its toes are parted, its
feet naked, its legs covered with feathers as far as
the heel, the bill fafhioned into a curved cone,
the tail long, and the head without comb or
membrane. A perfon who fticks rigidly to a
fyftem, could not fail to range it with the Peacock or the Pheafant, fince it poffeffes all the at-

---

* Edwards never fays that this bird difplays its tail: I may
therefore infer the negative, fince if the expanfion had taken
place, that intelligent naturalift would have obferved it and have
mentioned it.

† *Eperonnier,* from *Eperon,* a fpur.

tributes

tributes of that genus; but muft the hiftorian, exempt from prejudice and unfettered by forms, recognife it as the Peacock of Nature?

In vain it will be urged, that fince the principal characters of this bird are the fame with thofe of the Pheafant, the little variations ought not to feclude it from that arrangement; for I may ftill afk, who has a right to fix thefe principal characters? to decide, for inftance, that the negative attribute of having neither creft nor membrane is more effential than the fhape or the fize? and to pronounce that all birds which refemble each other in characters arbitrarily felected, muft alfo agree in their true properties?

In laying afide the name of Chinefe Peacock, I have acted conformably to the teftimony of travellers, who affure us, that through the whole extent of that vaft country there are no Peacocks but fuch as have been introduced from abroad*.

In this bird the iris is yellow, and alfo the fpace between the bottom of the bill and the eye; the upper mandible red; the lower mandible of a deep brown, and the feet of a dirty brown; its plumage is exceedingly beautiful; the tail, as I have already faid, is fprinkled with oval fpangles, and is of a fine purple colour with reflections of blue, green, and gold. The effect of thefe fpangles, or mirrors, is the more ftriking, as they

* Navarette, *Defcription de la China.*

are

are defined and diftinguifhed from the ground by a double circle, the one black and the other dull orange. Each quill of the tail has two of thefe mirrors cluftered together, the fhaft paffing between them. However, as the tail con-tains much fewer quills than that of the Peacock, it is much lefs loaded with fpangles; but to compenfate this, it has a very great number on its back and wings, where the Peacock has none: thofe on the wings are round; and as the ground colour of the plumage is brown, it refembles a fable richly ftrewed with fapphires, opals, eme-ralds, and topazes. The greater quills of the wing are not decorated with fpangles, all the reft have each only one; and their colours, whether in the wings or in the tail, do not penetrate to the other furface, which is of an uniform dull caft.

The male exceeds the fize of an ordinary Pheafant; and the female is a third fmaller, and appears more lively and active. As in the male, its iris is yellow; but there is no red on its bill, and its tail is much fmaller. And though in the female of this bird the colours are more like thofe of the male, than in the Peacocks or Pheafants, they are more faint and dull, and have not that luftre and thofe luminous undulations which pro-duce fo charming an effect in the fpangles of the male.

This

N.º 42

THE MALE CURASSO.

This bird was alive laſt year at London, and Sir ⸺ Codrington ſent coloured drawings of it to the younger Daubenton, from which our figures were taken.   [A]

[A] Specific character of the *Pavo Bicalcaratus:* " It is " brown ; its head ſomewhat creſted ; two ſpurs."

The HOCCOS.

ALL the birds known under the general term Hocco, are ſtrangers to Europe ; they belong to the warmer parts of America.   And the various names beſtowed by different Savages, each in his own jargon, have contributed, no leſs than the multiplied epithets impoſed by nomenclators, to introduce confuſion.   I ſhall endeavour, as far as the poverty of obſervation will permit, to diſpel the chaos, and reduce the nominal to real ſpecies.

I.

The HOCCO, *properly ſo called.*

Crax-Alector, Linn. and Gmel.
Crax Guianenſis, Briſſ.
Mituporanga, Ray.
The Indian Cock, Pitfield.
The Peacock Pheaſant of Guiana, Bancr.
The Creſted Curaſſow, Lath. Brown, and Sloane.

Under this ſpecies I range not only the *Mitou* and the *Mitou-poranga* of Marcgrave,

which

which that author confidered as of the fame
kind, the Indian cock of the Academicians and
of many others, the *Mutou* or *Moytou* of Laët,
the *Temocholli* of the Mexicans, and their *Tepetotoil*
or mountain-bird, the *Quirizao* or *Curaſſo* of Jamaica, the *Pocs* of Friſch, the *Hocco* of Cayenne
in Barrere's ſyſtem, the *Hocco* of Guiana, or the
twelfth Pheaſant in Briſſon's ; but I alſo refer to
the fame diviſion, as varieties, the *Hocco* cf Brazil,
and even Albin's Red Hen of Peru*, or Briſſon's
eleventh ſpecies of Pheaſant, the Hocco of Peru,
the *Coxoliſli* of Fernandez, and ſixteenth Pheaſant in Briſſon's ſyſtem. My reaſon for this
arrangement is, that this multitude of names is
applied to birds having many common characters, diſtinguiſhed only by ſome ſlight variations in the diſpoſition of the colours, in the
faſhion of the bill, and in ſome other circumſtances, which, in the fame ſpecies, are affected
by the age, ſex, and climate ; and theſe diver-
ſities are the more to be expected in a ſpecies
like the preſent, which is ſo eaſily tamed, and
has actually been tamed in ſome provinces, and
conſequently muſt partake, in ſome degree, of
the changes to which domeſtic animals are ſubject †.

* *Albin.* " It is of the fame ſize and figure with the Caraſſow
" hen, and appears to be of the fame ſpecies." Thus ſpeaks
Albin, who had the advantage of delineating the two birds from
the life.

† Sir Hans Sloane ſays, that their plumage is variegated in
different ways, like that of common hens.

The

The Academicians had heard that their Indian cock was brought from Africa, where it was called *Ano*; but as Marcgrave and feveral other obfervers inform us that it is a native of Brazil, and fince we learn from a comparifon of the moft accurate defcriptions and figures, that its wings are fhort and its flight laborious, we can hardly be perfuaded that it could traverfe the immenfe ftretch of ocean that divides the fhores of Africa and Brazil. It is much more natural to fuppofe that the fubjects diffected by the Academicians, if they were really brought from Africa, had been previoufly carried thither either from Brazil or from fome other fettlement in the New World. The fame reafon will enable us to judge of the propriety of the appellation of the Perfian Cock, beftowed by Johnfton on this bird.

The Hocco is nearly as large as a turkey. One of its moft diftinguifhing properties is a creft, which is black, or fometimes black mixed with white, about two or three inches high, and which extends from the origin of the bill to the back of the head. The bird can raife or deprefs it at pleafure, and according as it is differently affected. This creft confifts of narrow tapering feathers fomewhat reclined, but the point is reflected and bent forwards. Of thefe feathers, the Academicians obferved many whofe webs were inclofed half the length of their fhaft, in a kind of membranous cafe.

The

The prevailing colour of the plumage is black, which is for the moft part uniform and like velvet on the head and neck, and fometimes fprinkled with white fpeckles; the reft of the body has greenifh reflections, and in fome fubjects it changes into a deep chefnut, as in that of No. 125. of the *Plancves Enluminées*. The bird figured in that plate has no white under the belly or on the tail; in which refpect it differs from that of No. 86. Laftly, Others are white below the belly and not at the tail, and *vice verfâ;* and we muft obferve, that thefe colours are liable to vary both in their tints and in difpofition, according to the fex.

The bill is fhaped like that of the gallinaceous tribe, but is rather ftronger: in fome, it is of a flefh colour and whitifh near the point, as in the Brazilian Hocco of Briffon: in others, the end of the upper mandible is grooved on both fides, which makes it look as if it were armed with three points, the principal one in the middle, and the two lateral, formed by furrows, fomewhat farther back, as in the Indian Cocks of the Academicians: in others, the bafe is covered with a yellow fkin, in which are placed the noftrils; as in the Guiana of Briffon: in others, this yellow fkin, extending on both fides of the head, forms a circle of the fame colour round the eyes; as in the *Mitou-poranga* of Marcgrave: in others, this fkin fwells on the bafe of the upper mandible into a kind of tubercle or round bump, which

THE FEMALE CURASSO.

which is pretty hard and about the fize of a
fmall nut. It is commonly believed that the
females have not this protuberance; and Edwards
adds, that it does not grow on the males till after
the firft year; which appears to be the more
probable, fince Fernandez obferved in his *Tepetototl* a kind of tumor begun to form on the bill.
Some individuals, as the *Mitou* of Marcgrave, have
a white fkin behind the ear like the common
hens. The legs would refemble in fhape thofe
of the gallinaceous tribe, if they had fpurs and
were not proportionally thicker: they vary too
in their colour, from a darkifh brown to a car-
nation.

Some naturalifts would refer the Hocco to
the genus of the turkey; but it is eafy, from the
foregoing defcription, and from the figure, to
collect numerous and decifive differences which
difcriminate thefe kinds. In the turkey the
head is fmall and not feathered, which is alfo
the cafe with the top of the neck; the bill bears
a conical mufcular protuberance, capable of being dilated and contracted; the legs are armed
with fpurs; the tail feathers can be fpread like
a fan, &c. whereas in the Hocco, the head is
large, the neck funk, and both are clothed with
feathers; on the bill is a round, hard, and almoft
bony fwelling; and on the crown of the head
a moveable creft, which feems to be peculiar to
this bird, and which is raifed and depreffed at
                              pleafure;

pleafure; but no perfon has ever afferted that it can expand its tail feathers like a fan.

To thefe exterior differences, add the more intimate effential difparities, which appear from diffection to be as numerous.

The inteftinal canal is much longer, and the two *cæca* much fhorter, than in the turkey; its craw is alfo much lefs capacious, being only four inches round; but I have feen a crop taken out of a turkey, that feemed to have nothing unufual in its ftructure, that could contain half a Paris pint dry meafure. Befides, in the Hocco, the flefhy fubftance of the gizzard is for the moft part very thin, and its inner coat, on the contrary, very thick, and fo hard as even to be apt to crack. Laftly, The *trachea arteria* dilates and makes an inflexure near where it forks; as happens in fome aquatic birds, quite contrary to what is obferved in the turkey.

But if the Hocco be not a turkey, the modern nomenclators had ftill lefs reafon to fuppofe it a Pheafant; for befides thofe differences, which will readily be perceived externally and internally, there is a decifive one in the inftincts of thefe birds. The Pheafant is always wild; though bred from its infancy, though treated kindly and fed with great attention, it never becomes reconciled to the domeftic ftate; it is ever a reftlefs prifoner, ever feeking the means of efcape: it even abufes the companions of its flavery, and never affociates with them. When

it

it recovers its liberty, and is reſtored to the ſavage
ſtate, for which it ſeems to be formed, nothing
can be more timorous or miſtruſtful; every new
objeċt is viewed with a ſuſpicious aſpeċt; the
leaſt noiſe ſcares it, and the ſlighteſt motion
diſturbs its quiet; even the ſhadow of a branch
ſhaken by the wind is ſufficient to make it take
wing. On the contrary, the Hocco is a calm
bird, ſecure and even ſtupid; which perceives
no danger, or at leaſt makes no exertion to
ſhun it: it ſeems to forget itſelf, and to be
careleſs of its own exiſtence. Aublet ſhot nine
of them in the ſame flock with the ſame piece,
which he loaded as often as required. Such was
their patient tranquillity. We may ſuppoſe that
ſuch a bird muſt be ſociable; that it will readily
accommodate itſelf to the other domeſtic fowls;
and that it can be eaſily tamed. And though
trained, it roams to a great diſtance during the
day, but always returns again in the evening;
as Aublet tells me himſelf. It becomes ſo tame
as to rub with its bill on the door to gain ad-
miſſion; to pull the ſervants by the clothes
when they negleċt it; to follow its maſter
every where; or, if not allowed, it waits
anxiouſly for his return, and, on ſeeing him
again, ſhews every ſign of joy and affeċtion.

It is difficult to conceive habits more oppoſite;
and I ſhould imagine that no naturaliſt, or even
nomenclator, if he had been acquainted with
them,

them, would have ventured to refer thefe two
birds to the fame genus.

The Hocco loves to inhabit the mountains, if
we may infer this from the import of the name
Tepetototl, which, in the Mexican language, fignifies *mountain-bird*.   When kept in cages it is
fed on bread, pafte, and other fuch things *.   It
is fond of perching on trees, efpecially to pafs
the night.   It flies tardily, as I have obferved
above; but its carriage is bold †.   Its flefh is
white, though rather dry; but when kept a
fufficient time, it is pleafant eating ‡.

Sir Hans Sloane fays, that its tail is only two
inches long, which Edwards conceives to be
printed by miftake for ten   But I fhould imagine
that this correction is too general and unlimited;
for I obferve that Aldrovandus afferts, from a
drawing of a bird of this fort, that it has no tail.
And on the other hand, Barrere relates, from
his own obfervations which he made on the fpot,
that the female of his Amazon Hocco, which
is the Curaffow-Hocco of Briffon, has a fhort tail.
Whence it appears that what Sir Hans Sloane has
affirmed with regard to the Hocco in general,
muft be reftricted to the female only, at leaft
in certain tribes.   [A]

* Fernandez.   † Barrere.   ‡ Fernandez, Marcgrave, &c.

[A] Specific character of the *Crax-Alector:* " Its cere is yel-
" low, its body black, its belly white."

II. The

II.

The PAUXI, or STONE.

Crax-Pauxi, Linn. and Gmel.
Gallina Indica Alba, Will.
Crax Mexicanus, Briff.
The Cuſhew Curaſſow, Lath. and Edw.

We have figured this bird in the *Planches En-luminées* under the name of *Stone of Cayenne,* which is really what it bears in the *Royal Menagerie,* where the drawing was made after the life. But as in its native country, which is Mexico, it is known by the name of *Pauxi,* according to Fernandez, I have thought proper to employ both theſe names.—It is the fourth ſpecies of the Pheaſant of Briſſon, which he terms the *Mexican Hocco.*

This bird reſembles the preceding in many reſpects ; but it differs in ſome particulars. Its head is not tufted like the other ; the ſwelling on the bill is larger, of the ſhape of a pear, and of a blue colour. Fernandez ſays, that this tubercle is as hard as a ſtone ; and this is the reaſon, I ſuppoſe, why it was called firſt the *Stone-bird,* and then the *Stone ;* for the ſame cauſe that it was firſt named *Cuſco* or *Cuſhew-bird,* and *Numidian-hen,* from this bump, which ſome have conceived to reſemble the American nut, called *cuſco* or *cuſhew ;* and others have imagined that it is like the caſque of the Pintado.

7

But

But thefe are not the only differences which diftinguifh the Páuxi from the preceding Hoccos: it is fmaller, its bill is ftronger, more hooked, and almoft as much fo as that of the parrot. Befides, it is much more feldom brought to Europe than the Hocco. Edwards, who faw the Hocco in almoft every collection, could not meet with a fingle Cafhew or Pauxi in the courfe of his inquiries.

The elegant black of its plumage has blue and purple reflections, which cannot be reprefented in the defign.

This bird perches on trees; but it lays on the ground like the pheafants, leads its young, and even calls them together. The brood live firft on infects, and afterwards, when they are grown up, they feed on fruits, feeds, and whatever is proper for poultry *.

The Pauxi is as gentle, or, if we chufe, as ftupid, as the other Hoccos; for it will fit ftill though fired at fix times in fucceffion; yet, according to Fernandez, it will not fuffer itfelf to be caught or handled; and M. Aublet informs me, that it is found in uninhabited places, which is probably one of the caufes why it is fo rarely brought to Europe.

Briffon fays, that the male differs from the female only by the colours, having brown where the other is black; but that they are in other refpects alike. Aldrovandus, however, admit-

* Aublet and Fernandez.

ting

ting that the plumage is in general brown, obferves that its wings and tail are cinereous; that
the bill is lefs hooked, and that it has no tail;
which would be a feature of coincidence with
the Amazon Hocco of Barrere, in which, as we
have already noticed, the tail of the female is
much fhorter than that of the male: and thefe
are not the only American birds which want the
tail; in a certain part of that continent, the
poultry tranfported from Europe lofe their tail
and rump, as we have already obferved in the
hiftory of the cock. [A]

[A] Specific character of the *Crax Pauxi*: " Its cere is blue;
" a crefted bunch on its noftrils; its body blackifh; its belly and
" the tip of its tail, white."

III.

The HOAZIN.

Phafianus Criftatus, Gmel.
Crax Fufcus Mexicanus, Briff.
The Crefted Pheafant, Lath.

This bird is figured in the *Planches Enluminées*
under the name of *Crefted Cayenne Pheafant*; at
leaft it does not differ fenfibly from that, as will
appear by comparing No. 337. with the defcription of Fernandez.

According to that author, the Hoazin is not
quite fo large as a turkey-hen; its bill is hooked;
its breaft of a yellowifh white; the wings and
tail marked with fpots or white rays an inch
afunder; the back, the upper fide of the neck,
the fides of the head, are of a tawny brown; the
legs are of a dirty colour. It has a creft compofed of feathers that are whitifh on one fide
and black on the other; this creft is taller and
differently fhaped from that of the Hoccos; and
it does not appear that they can raife and deprefs
it at pleafure: its head alfo is fmaller and its
neck more flender.

Its voice is very ftrong, and more like a bray-
ing than a cry. It is faid that it calls its own
name, probably in a fad frightful tone. Nothing
more was wanted among favage tribes to place
it in the clafs of inaufpicious birds; and as the
human mind is naturally prone to imagine the
object of dread endowed with vaft power, thefe
rude people draw from it remedies for the moft
inveterate and alarming diforders. They do not
appear, however, to feed on it: they abftain perhaps through fear, which it infpires; or perhaps from an averfion, becaufe it lives commonly on ferpents. It inhabits generally extenfive forefts, where it perches on trees befide
water, to watch and furprife thefe reptiles.
It is found in the warmeft parts of Mexico.
Hernandez adds, that it appears in autumn,
which

which gives room to fufpect that it is a bird of paffage *.

M. Aublet affures me that this bird, which he eafily recognized in No. 337. of the *Planches Enluminées*, can be tamed; and that it is fometimes a fort of domeftic among the Indians, and that the French call it a peacock. The young are fed with ants, worms, and other infects. [A]

* Hernandez.—Fernandez fpeaks of another bird to which he gives the name of *Hoazin*; though from his account it appears to be very different from what we have defcribed; for befides that it is fmaller, its cry is very agreeable, and refembles fometimes a laugh or a fneering laugh : its flefh is eaten, though neither tender nor well tafted.—The bird cannot be tamed.

I fhould rather difcover the Hoazin in another bird mentioned by the fame author, after the Pauxi. He thus defcribes it: " Another bird muft be ranged with the Pauxi. . . . It is of the " fize of a ftork, its colour cinereous, the creft eight inches long, " and compofed of many feathers . . . thefe dilated, efpecially on " the top." Here is diftinctly the creft and the fize of the Hoazin.

[A] Specific character of the *Phafianus Criftatus:* " Above, " brown; below, rufous-white; its vent rufous, its head crefted, a " naked red fpace about the eyes; the tail wedge-fhaped, with " a yellow tip."

IV. The

IV.

The YACOU.

Penelope Criſtata, Gmel.
Meleagris Criſtata, Linn.
Phaſianus Fuſcus Braſilienſis, Klein.
Iacupema, Marcg. Ray, and Will.
The Guan, or *Quan,* Edw. and Lath.

This bird has named itſelf; for its cry is, according to Marcgrave, *Yacou;* whence is derived the name *Iacupema.* I have preferred that of Yacou as the eaſieſt, and the beſt adapted.

Marcgrave is the firſt who has ſpoken of this bird. Some naturaliſts, copying him, have ranged it with the pheaſants; others, ſuch as Briſſon and Edwards, have claſſed it with the turkies. But it is neither the one nor the other:—it is not a turkey, though it has a red ſkin under the neck; for it differs in many reſpects; in its ſize, which is ſcarcely equal to that of a common hen; its head is partly covered with feathers, and its creſt is much more like that of the Hoccos than that of the creſted turkey; and its legs have no ſpurs:—beſides, it has not the bunch of hard hair under the neck, nor the muſcular caruncle on the bill, as in the turkey-cock, nor does it expand the feathers of its tail. On the other hand, it is not a pheaſant; for it has the long and ſlender bill and the creſt of

the

the Hoccos; its neck is flender; it has a flefhy membrane under the throat; its tail-feathers are all of an equal length; and its difpofitions are mild and gentle: all which characters diftinguifh it from the pheafants, and its cry differs from both that of the pheafant and of the turkey. But what fhall we then make it? It fhall be a Yacou, having fome analogies with the turkey (the flefhy membrane under the throat and the tail compofed of equal quills); with the pheafants (the eye encircled with black fkin, the wings fhort, and the tail long); with the Hocco (the long tail, the creft, and mild difpofition); but which is diftinguifhed from all thefe by numerous and marked differences, and therefore conftitutes a feparate fpecies.

We can hardly doubt that the *Guan* or *Quan* of Edwards, fo called, according to him, in the Weft Indies, probably by fome other tribe of Savages, is at leaft a variety of the Yacou, from which it differs only in being not fo tall, and its eyes of another colour, but fuch differences may take place in the fame fpecies, efpecially fince it is domefticated.

Black mixed with brown is the prevailing colour of its plumage, but with different reflections, and fome white ftreaks on the neck, breaft, belly, &c.; the legs are of a bright red.

The flefh of the Yacou is excellent meat. All that is known with refpect to its other properties has been related in the beginning of this article.

Ray

Ray confiders it as of the fame fpecies with the coxolitli of Fernandez; but that bird is much larger, and has not under its throat that flefhy membrane which characterizes the Yacou; and for this reafon I have claffed it with the Hoccos properly fo called.　[A]

[A] Specific character of the *Penelope Criftata*: "Its head is "crefted with erect feathers, its temples violet."

V.

The MARAIL.

Penelope-Marail, Gmel.
The Marail Turkey, Lath.

No author has taken notice of the female of the Yacou except Edwards, who conjectures that it has no creft. From this fingle authority, and the comparifon of the moft accurate figures and ftuffed fpecimens, I am inclined to fuppofe, that the bird figured in No. 328. of the *Planches Enluminées* under the name of *The Greenifh Pheafant of Cayenne*, and which is generally called in that ifland *The Marail*, is perhaps the female of the fpecies of Yacou; for I can difcover many decifive points of refemblance to the *Guan* of Edwards (Plate XIII.); in its fize, the colour of its
plumage,

plumage, and its general fhape, if we except only the creft, which is wanting in the female; in its port, in the length of its tail, in the red circle that furrounds the eyes *, the red naked fpace below the throat, the form of its feet and bill, &c. I muft own that I have alfo perceived fome differences; the quills of the tail are like organ pipes, as in the pheafant, and not equal, as in the Guan of Edwards; and the noftrils are not fo near the origin of the bill: but it would not be difficult to inftance a number of fpecies in which the female differs ftill more from the male, and in which there are varieties that are more remote from each other.

M. Aublet, who faw this bird in its native country, tells me, that it is eafily tamed, and that its flefh is delicate, and richer and fuperior in fucculency to that of the pheafant. He adds, that it is a real turkey, only fmaller than what is naturalized in Europe: and this is ftill another point of refemblance to the Yacou, its having been taken for a turkey.

This bird is not only found in Cayenne; but, if we may judge from the identity of the name, it inhabits the country which is watered by the majeftic ftream of the Amazons; for Barrere fpeaks of the Marail of the Amazons, as a bird whofe plumage is black, its bill green, and no

---

* This naked fkin is blue in the Yacou, and red in the Marail; but we have before obferved the fame difference of colour between one fex and the other in the flefhy membranes of the Pintado.

tail.

tail. We have feen, in the account of the
Hocco and the Stone of Cayenne, that in thefe
fpecies fome individuals without tails have
been taken for females: is this the cafe too with
the Marails? With regard to moft of thefe
foreign birds fo little known, if we adhere to
veracity, we muft fpeak with diffidence and
hefitation. [A]

[A] Specific character of the *Penelope Marail:* – " It is
" greenifh black, the fpace about its eyes naked, and its feet
" red; the throat fomewhat naked, and dotted with white."

VI.
The CARACARA.

I give this name, which is expreffive of its
cry, to that beautiful bird of the Antilles defcribed by Father du Tertre.

If all the American birds that have been taken
for pheafants muft be referred to the Hocco
tribe, the Caracara ought to be ranged with
thefe; for the French inhabitants of the Antilles,
and Father du Tertre after them, have applied
to it the name of *Pheafant.* " This Pheafant is,"
fays he, " a very beautiful bird, about the fize
" of a capon, taller, and with legs like thofe of
" the peacock. Its neck is much longer than
" that of a cock, and the bill and head refemble
" thofe of a raven; all the feathers of the neck
" and

" and breaſt are of a fine ſhining blue, as plea-
" ſing as peacock's plumes; all the back is of a
" brown gray, and the wings and tail, which
" are rather ſhort, are black."

" When this bird is tamed, it becomes maſter
" of the houſe, and drives off the common hens
" and turkey-hens, and ſometimes even kills
" them. Nor does it allow the dogs to paſs
" without offering violence. . . . . I ſaw one
" which was a mortal enemy to Negroes, and
" would not permit one to enter the hut, but
" picked their legs and feet ſo cruelly as to draw
" blood." Thoſe who have eaten them affirm,
that their fleſh is as good as that of the pheaſants in France.

How could Ray ſuppoſe that ſuch a bird was
the ravenous bird mentioned by Marcgrave under the ſame name? It is true indeed that it
fights with the poultry, and flies at dogs and
Negroes; but this it does only when tamed.
We ſhall more eaſily diſcover in it the natural
jealouſy of a domeſtic animal, which cannot
bear the rivals in his maſter's favour, than the
ferocious diſpoſitions of a bird of prey, which
darts on others to tear them in pieces and devour
them. Beſides, it is not common that the fleſh
of a rapacious bird is delicate eating, as is that
of the Caracara. Laſtly, It appears, that in the
Caracara of Marcgrave, the tail and wings are
much longer in proportion than in that of Father du Tertre.

VII. The

VII.

The CHACAMEL.

Penelope Vociferans, Gmel.
The Crying Curaſſow, Lath.

Fernandez ſpeaks of a bird which is of the ſame country and nearly the ſame ſize with the preceding, and which, in the Mexican language, is called *Chachalacamelt;* from which I have formed *Chacamel*, for the eaſier pronunciation. Its chief character is that of having a cry like the common hen, or rather like the clamorous noiſe of a number of fowls; for it is ſo conſtant and ſo loud, that a ſingle bird of this kind is ſaid to make as much din as a whole court-yard. Hence is derived the Mexican name, which ſignifies the *crying bird*. It is brown on the back, of a duſky-white on the belly, and the bill and feet are bluiſh.

The Chacamel, like moſt of the Hoccos, commonly inhabits the mountains, where it rears its young. [A]

[A] Specific character of the *Penelope Vociferans* :—" Its bill " is bluiſh, its back brown, its breaſt blue, its belly whitiſh- " brown."

VIII. The

VIII.

The PARRAKA and HOITLALLOTL.

Phasianus Parraqua, Lath. Ind.
The Parraka Pheasant, Lath. Syn.
Phasianus Mexicanus, Gmel.
The Courier Pheasant, Lath.

As far as we can judge from the imperfect hints of Fernandez and Barrere, we may range here, 1. The Parraka * of the latter, which he calls *Pheasant,* and of which he says only that the feathers of the head are of a tawny colour, and form a kind of creft. 2. The *Hoitlallotl,* or Long Bird of the former, which inhabits the warm regions of Mexico †. This bird has a long tail, short wings, and a laborious flight, like most of the foregoing; but it outstrips the fleetest horses. It is not so large as the Hoccos, being only eighteen inches from the tip of the bill to the end of the tail. Its general colour is white, verging on the fulvous. Near the tail it is stained with black, mixed with some white spots; but the tail itself is of a varying green, which has reflections nearly like the peacock's plumes.

---

* Specific character :—" It is brown, below fulvous, its top
" fulvous, its tail equal."

† Specific character :—" It is tawny-white, its tail long and
" green."

These

Thefe birds are fo little known, that we cannot venture to refer them to their fpecies. I range them here only becaufe thofe few properties which we do know belong more to the birds juft defcribed than to others. Their true place muft be affigned from actual obfervation. In the mean time, I have done what I can to draw the curiofity of thofe who have it in their power to obferve the facts.

M

The PARTRIDGE.

IT is often the moſt difficult to give an accurate and conſiſtent account of thoſe ſpecies
which are the moſt generally known. When a
perſon meets, for the firſt time, with a bird
which he has never before ſeen, he overlooks
the minute characters, and, ſeizing the more
obvious reſemblances, he refers it to that tribe
with which he is previouſly beſt acquainted.
Hence that ſtrange incoherent jumble of names
which have been formed on the relations of
haſty and inaccurate obſervers. We have already been more than once embarraſſed in this
chaos of terms; and I am afraid that the article
of the Partridge will not be the laſt which requires a critical examination.

I take the Common Partridge for the baſis and
firſt ſpecies of the genus, as being the beſt known,
and therefore the fitteſt ſubject for compariſon.—I ſhall admit one variety and three permanent breeds.

Theſe permanent breeds are, 1. The *Common
Gray Partridge;* and, as a variety of it, the
White Gray Partridge of Briſſon. 2. The *Da-
maſcus Partridge,* not that of Belon, which is
the Hazel Grous, but that of Aldrovandus, which
is ſmaller than our Gray Partridge, and which
appears

appears to be the fame with the Little Partridge, a bird of paffage well known to our fportfmen. 3. The *Mountain Partridge*, which is figured in No. 136. of the *Planches Enluminées*, and which feems to form the fhade between the Gray and the Red Partridges.

In the fecond fpecies I range the Red Partridge, into which I admit two permanent breeds in France, and one variety, and two foreign breeds.

The two permanent Red Partridges which are found in France are, 1. No. 150. *Planches Enluminées*. 2. The *Greek Partridge*, Pl. 231.

The two foreign fpecies are, 1. The *Red Barbary Partridge* of Edwards, Pl. 70. 2. The *Rufous-breafted Partridge*, which is found on the banks of the Gambra.

As the plumage of the Red Partridge is liable to affume fhades of white like that of the Gray Partridge, thence refults a variety exactly fimilar to that in the latter.

From this genus I exclude feveral fpecies which have been improperly referred to it.

1. The Francolin, which we have removed from the Partridges, becaufe it differs from them not only by its general fhape, but by fome particular characters, as in the fpurs, &c.

2. The bird called by Briffon the *Senegal Partridge*, and which he makes his eighth fpecies. This bird appears to me to be more a-kin to the Francolins than to the Partridges;

7   and

and as it has two fpurs on each leg, I fhall give it the epithet of *Double Spur*.

3. The African Red Partridge.

4. The third foreign fpecies, called by Briffon the *Great Partridge of Brazil*, which he fuppofes to be the *Macucagua* of Marcgrave, from whom he copies the defcription, and confounds it with the *Agamia* from Cayenne, which is a bird wholly different from both.

5. The *Yambou* of Marcgrave, which is the Brazilian Partridge of Briffon, and which has neither the fhape, the habits, nor the characters of the Partridge; fince, according to Briffon himfelf, it has a long bill, perches upon trees, and lays blue eggs.

6. The American Partridge of Catefby and Briffon, which alfo perches, and prefers the woods to the cleared grounds; a character which does not belong to the Partridge.

7. A multitude of American birds, which the herd of travellers have called Partridges from fome flight refemblance inaccurately obferved. Such are the birds known at Guadeloupe under the names of *Red Partridges*, *Gray Partridges*, and *Black Partridges*; though, according to the accounts of perfons better informed, they are pigeons or turtles; fince they have neither the bill nor the flefh of the Partridge, perch on trees, where they build their nefts, lay only two eggs, and fince the young do not run as foon as they quit the fhell, but are fed by the parents in the

nest

neft like turtles. Such too are moft probably thofe Partridges which Carreri faw on the mountains of the Havannah; fuch the *Manbouris*, the *Pegaſſous*, and the *Pegacans* of Lery; and fuch perhaps are fome American birds which I have ranked in the clafs of Partridges on the authority of writers, when their relations feemed not contradicted by facts; though I muft own, that it is not likely birds fo heavy could crofs the intervening ocean.

The GRAY PARTRIDGE*.

Tetrao-Perdix, Linn. and Gmel.
Perdix Cinerea, Ray, Will. and Briſſ.
The Common Partridge, Penn. and Lath.

Though Aldrovandus, judging of other countries from his own, afferts that Gray Partridges abound in every part of the globe; it is certain that there are none in the ifland of Crete; and it is probable that they never inhabited Greece; for Athenæus remarks with furprize, that all the Italian Partridges had not a red bill like thofe of Greece. Nor are they

* In Italian, *Perdice*; in Spaniſh, *Perdiz*; in German, *Wild-bun*, or *Feld-bun*; in Swediſh, *Rapp-boena*; in Poliſh, *Kuroptwa*.

equally

equally spread through Europe : they seem to
avoid the extremes both of heat and cold, and
are found neither in Africa nor in Lapland.
They thrive most in the temperate parts of
France and Germany. It is true, indeed, that
Boterius says that they do not inhabit Ireland ;
but this must be understood of the Red Partridges,
which are not found even in England, (according to the best authors of that country,) and
which have not penetrated in that direction beyond the islands of Guernsey and Jersey. The
Common Partridge is frequent in Sweden, where
Linnæus tells us it winters under the snow in a
kind of burrow, which has a double entrance.
This manner of lodging during the severe season, is very like that of the Ptarmigan, which
we have already described ; and if this fact were
not averred by a man of so high reputation as
Linnæus, I should suspect some mistake ; especially as in France the long winters, with great
falls of snow, prove fatal to numbers of Partridges. Lastly, as it is a bird of laborious
flight, I am much inclined to suspect that it has
never migrated into America ; and I should
imagine, that those birds of the New World
which are referred to this genus, would be separated from it if they were better known.

The Gray Partridge differs in many respects
from the Red ; but what makes me consider
them as distinct kinds, is that, according to the

remark of the few fportfmen who can make obfervations, though they fometimes inhabit the
fame fpot, they never affociate together. A
vacant male of the one fpecies will fometimes,
indeed, confort with a pair of the other, follow
them, and even difcover fymptoms of jealoufy;
yet it never copulates with the female, though
it is reduced to abftinence, and beholds continually the fweets of conjugal felicity, and feels
the enlivening influence of fpring.

The Gray Partridge is alfo of a gentler nature than the Red, and not difficult to tame;
and when not teazed, it foon becomes familiar *. However, they never could be formed into flocks that would be driven, as has
been done with the Red Partridges: for the
Red Partridges are thofe which travellers, as
Olina remarks, defcribe as being bred in numerous flocks on fome iflands of the Mediterranean. The Gray Partridges have alfo a
more focial turn, fince each family continues in
a fingle body, or covey, till the pairing of love.
If a hatch, from fome accident, does not completely fucceed, the families recruit their ftrength
by uniting with others, and adopting the feeble
remnants of fuch as have fuffered moft feverely
from the fportfmen: fo that about the end of

* Ray afferts the contrary; but as he confeffes that the Red
Partridges are never feen in England, his authority will not in
this inftance weigh againft actual obfervers.

fummer

summer they often compose new coveys more numerous than at first, and which continue associated till next year.

These birds are fond of corn countries, especially where the fields are in high cultivation, and manured with marl; no doubt because they find there abundance of food, both grain and insects; and perhaps the saline quality of the marl, which contributes so much to the fertility of the soil, is also suited to their constitution or taste. Gray Partridges prefer the open country, and never resort to copses or vineyards, but when they are pursued by the fowler, or by the bird of prey: yet they do not lodge in the depths of the forests; and I have been frequently told that they never pass the night among bushes or thickets: however, a Partridge's nest was found in a bush at the root of a vine. They begin about the end of winter, after the intense frosts, to pair: that is, each male selects his female companion, and retires. But this new arrangement is not effected without violent disputes among the males, and sometimes even among the females. War and love are in most animals inseparable, especially among those which, like the Partridges, are stimulated by an ardent appetite. The females of this species, like the common hens, lay without having had intercourse with the male. When the Partridges are once paired, they never part, but live in the closest and the

most

moſt faithful union. Sometimes, after they are paired, the weather grows ſevere, and then they all gather together, and again form the covey.

Gray Partridges ſeldom breed, at leaſt in France, before the end of March, above a month after they have begun to pair; and they do not lay before May, or even June, if the winter has laſted long. They make their neſt, in general, with little care or preparation: ſome graſs or ſtraw, ſtrewed roughly in the print of an ox or a horſe's foot, is all they require. It is obſerved, however, that the older and more experienced females take greater pains with their neſts than young ones, and are more careful both in guarding againſt inundation, and in chuſing a ſpot ſomewhat elevated and pro-tected naturally by bruſh-wood. They gene-rally lay from fifteen to twenty eggs, and ſome-times twenty five; but the number is much ſmaller when the bird is either very young or very old: ſuch too is the ſecond hatch made by Partridges of the proper age, when the firſt has been deſtroyed. The eggs are nearly of the ſame colour with thoſe of pigeons: Pliny ſays that they are white*. The incubation laſts about three weeks, more or leſs according to the degree of heat of the ſeaſon.

The female takes upon herſelf the whole taſk of covering, and, during that time, ſhe under-

* Lib. x. 3.

goes

goes a confiderable moult; for all the feathers
of the belly drop. She fits with great affiduity;
and, it is faid, that fhe never leaves her eggs
without ftrewing them with leaves. The male,
attentive to his mate, generally fettles near the
neft, ready to accompany her when fhe rifes in
queft of food; and his attachment is fo faithful
and fteady, that he prefers this laborious office
to the free pleafures which the calls of other
Partridges folicit him to enjoy: to thefe he
fometimes replies, but never quits his ftation to
indulge his appetite. At the expiration of the
regular time, if the feafon be favourable and the
incubation fucceed, the chicks pierce the fhell
with great eafe, and as foon as they have extri-
cated themfelves, they begin to run, carrying
fometimes a part of the fhell with them. It
happens fometimes, however, that they are unable to burft from their prifon, and that they
die in the ftruggle. In this cafe, the feathers of
the young bird are found glued to the inner
furface of the fhell, which muft happen whenever the egg is expofed to too great heat. To
remedy this malady, dip the eggs five or fix
minutes in water, fo that the moifture may foak
through the fhell and loofen the feathers. This
kind of bathing may alfo perhaps refresh the
young bird, and give it additional ftrength to
force a paffage. The fame happens with regard
to pigeons, and many other ufeful birds, which

might

might be faved by the method I have defcribed, or fome analogous experiment.

The male, though it has no fhare in the incubation, affifts the mother in raifing the young. They lead them in common, continually call them together, point out to them their proper food, and teach them to find it by fcratching the ground with their nails. It is not uncommon to difcover them fquatted befide each other, covering the chickens with their wings, whofe heads project on all fides, prefenting very lively eyes. In fuch cafe, the parents are not eafily flufhed; and the fportfman, who is attentive to the prefervation of his game, avoids difturbing fo interefting an office. But if the pointer comes too near, or runs in upon them, the male is always the firft that fprings, venting his anguifh in a peculiar cry, and appropriated to this emergence. He ftops thirty or forty paces diftant, and fometimes even he returns upon the dog and beats it with his wings,—to fuch a degree does parental affection infpire courage in the moft timid animals! Sometimes that tender fentiment infpires in thefe birds a fort of prudence, and fuggefts expedients for faving the brood. When the male fprings in fuch cafes, he has been obferved to fly flowly, and hanging his wing, as if to decoy the enemy into a purfuit, in the expectation of an eafy prey; while the bird keeps always before him, but at fuch a fhort diftance as continually

tinually to afford hopes, till the fportfman is carried away from the covey. On the other hand, the female fprings fhortly after the male, and fhoots to a much greater diftance, and invariably in a different direction. Immediately after fhe has alighted, fhe returns back running along the furrow, and finds her chickens fcattered and fquatted among the grafs and the leaves; haftily collects them, and before the dog has returned from the eager purfuit of the male, fhe has led them to a diftant fpot, without giving the flighteft notice to the fportfman by the noife in retreating. It is an obfervation with refpect to animals which holds very generally, that the ardor for copulation is the meafure of the folicitude for their progeny. The Partridge is an inftance: few birds are fo amorous, and few difcover fuch an affiduous or fuch bold vigilance for their young. This ftrength of affection fometimes degenerates into rancour, which the mother difcovers to other coveys, purfuing them and tearing them with her bill.

The legs of the young Partridges are at firft yellow, which colour grows lighter, running into white, and afterwards turns brown, and at laft, when the bird is three or four years old, it deepens into black. This is a method of difcovering their age: another is drawn from the appearance of the laft feather of the wing,

which

which is pointed after the firſt moult, but in the following year is quite round.

The firſt food of young Partridges is ants eggs, and the ſmall inſects which they find on the ground and among the herbage. Thoſe which are fed within doors refuſe grain for a long time, and probably this is not their proper aliment till they are grown up. They always prefer lettuce, ſuccory, chickweed, ſow-thiſtle, groundſel, and even the ſhoots of ſpringing corn. In the month of November their ſtomach is found filled with that ſubſtance, and during the winter they learn to ſearch for the tender herbage beneath the ſnow. But when the ground is ſtiff with froſt, they reſort to the mild ſprings, and crop the herbs on their margins, though not ſuited to their conſtitution.— In ſummer, they are never obſerved to drink.

Partridges are three months old when the red tint diſcovers itſelf; for the Gray Partridges alſo have red on the ſide of the temples between the eye and the ear, and its appearance is a critical period with theſe birds, as with all others in like caſes, ſince it advances the adult ſtate. Previous to this they are delicate, their wing feeble, and they dread the damps; but after they have recovered from the ſhock, they become hardy, and begin to ply their wings, to ſpring together and conſtantly keep company; and though the covey be diſperſed, they learn to aſſemble again, in ſpite of the precautions of the ſportſman.

4   They

They affemble by a call ; every body knows the cry of the Partridge, which is not very pleafant ; it is rather a fharp grating noife like that of a fcythe, than a warble. The mythologifts, ftruck with this refemblance, metamorphofed the inventor of that inftrument into a Partridge *. The note of the male differs not from the female, except in being louder and more drawling. The male is befides diftinguifhed from the female by a blunt fpur on each leg, and a black mark like a horfe-hoe under the belly, which is not found in the other fex.

In this fpecies, as in many others, there are more males hatched than females †; and it is a matter of fome confequence to deftroy the fupernumerary males, which difturb the pairs already formed and check multiplication. The common method of catching them is to call them in the pairing feafon by means of a female, and the beft for this purpofe is one that has been taken old. The males flock to the female's voice, and fall into the fowler's fnares. So headlong they rufh into danger, as fometimes to alight on houfes, or even on the fhoulder of the bird-catcher. The moft proper fort of fnares, and what are the leaft liable to accidents, are a kind of large weel nets of a tunnel fhape, into which the Partridges are driven by a perfon dif-

---

* Ovid's Metamorphofes, Book viii.
† About a third more, according to Ray.

guifed

guifed like a cow, who, to aid the deception, holds in his hand one of the bells ufually faftened to the necks of cattle *. After they are entangled in the lines, the fupernumerary males are felected, and fometimes even all the males are taken, and the females are fet at liberty.

The Gray Partridges are fedentary birds, which not only continue in the fame country, but which ftray as little as poffible from the neighbourhood where they are bred, and they always return again. They fear much the bird of prey; when they perceive him, they crowd clofe one upon another, but their formidable enemy difcovers them from a diftance, approaches them glancing along the ground, in order to fpring one of which he may catch on the wing. Surrounded by fo many enemies, and expofed to fo many dangers, we may naturally fuppofe that few will reach a great age. Some fix the period of their life at feven years, and affert that, in their fecond year, they have attained their full vigour, and give over laying in the fixth. Olina fays, that they live twelve or fifteen years.

It has been tried to breed Partridges in parks, for the purpofe of ftocking grounds not inhabited by them. The experiment has fucceeded, and it has been found that they may be raifed nearly the fame way as Pheafants, only no dependence muft be had on the eggs of domeftic

* Olina.

Partridges,

Partridges. Seldom do they lay, when reduced to
that ftate, and ftill feldomer do they pair and copulate; and they never have been obferved to
hatch in thofe inclofures where the Pheafants fo
readily breed. We muft therefore fearch in the
fields for the eggs of free Partridges, and fet
them under common hens. A fingle hen can
hatch about two dozen and rear that number of
young, which will follow this ftranger as well
as they would do their natural mother, but they
are not fo well acquainted with her voice. They
become however familiarized to it in a certain
degree, and the Partridge thus bred, retains
through the reft of its life the habit of calling
when fhe hears the clucking of hens.

The Gray Partridges are much lefs delicate to
raife than the Red fort, and not fo fubject to dif-
eafes, at leaft in France, which it would feem is
their congenial climate. It is unneceffary even
to give them ants eggs, and they may be fed
like the common poultry on bread crumbs, hard
eggs, &c. When they have grown ftout and
begin to feek their food, they may be let loofe
where they are bred, and from which place they
never, as I have already noticed, remove to any
great diftance.

The flefh of the Common Partridge has long
been efteemed delicious and wholefome food.
It has two properties which are feldom combined; it is juicy, and yet not fat. Thefe
birds have twenty-two quills in each wing, and
eighteen

eighteen in the tail, of which the four mid-ones are of the fame colour with the back *.

The noftrils, which are at the origin of the bill, are more than half covered with a fcreen of the fame colour with the bill, but of a fofter fubftance, as in the common hens. The naked fpace between the eye and the ear is of a brighter red in the male than the female.

The inteftinal canal is about two feet and a half long, and the two *cæca* are each five or fix inches. The craw is very fmall †, and the gizzard is full of gravel mixed with the food, as obferved in all the granivorous tribes. [A]

[A] Specific character of the Common Partridge, *Tetrao Perdix* :—" It has a naked faffron fpot under its eyes, its tail is " ferruginous, its breaft brown, its feet whitifh." Its egg is fomewhat pear-fhaped, and greenifh gray.

———————

The GRAY-WHITE PARTRIDGE.

Tetrao Perdix, Var. Linn
Perdix Cinereo Alba, Briff.

This bird was known to Ariftotle ‡, and noticed by Scaliger §; for they both fpeak of the

* Willughby.
† *Ingluvies ampla,* fays Willughby, but in the Partridge which I diffected it was very fmall.
‡ Lib. v. 6. § Exercit. 59. in Cardanum.

White

White Partridge, and we cannot fuppofe that they meant the Ptarmigan, fometimes improperly fo called; fince Ariftotle could not be acquainted with that bird, which is not an inhabitant of Greece, Afia, or any country to which his information reached.   And indeed what proves decidedly this pofition, he does not remark the characteriftic property of this bird, that its feet are feathered to the toes.   With regard to Scaliger, he could not poffibly confound thefe two fpecies; becaufe in the fame chapter where he fpeaks of the White Partridge of which he had eaten, he a little afterwards difcourfes at great length on the *Lagopus* of Pliny, whofe feet are clothed with plumage, and which is the true Ptarmigan.

The Grayifh White Partridge is by no means fo white as the Ptarmigan; the ground only is of that colour, and is fullied with the fame fpecks as in the Common Partridge, diftributed in the fame manner.   But what fully proves that the difference in the colour of the plumage is only accidental, and forms a variety of the Gray Partridge, is, that, according to naturalifts and even according to fportfmen, it mingles and affociates with that fpecies.   One of my friends * faw a covey of ten or twelve Partridges which were entirely white, and was witnefs to their pairing with the common fort in the breeding feafon.

* Le Roi, *Lieutenant des Chaffes* at Verfailles.

Thefe

Thefe White Partridges had white eyes or rather white pupils, as happens too in white hares, white mice, &c. the bill and legs were of a lead colour.

───────────

The DAMASCUS PARTRIDGE.

La Petite Perdrix Grife, Buff.
Tetrao Damafcenus, Gmel.
Perdix Damafcena, Briff. &c.

This Partridge, defcribed by Aldrovandus, is probably the fmall migratory fpecies, which has been obferved at different times in feveral provinces of France.

It differs from the Gray Partridge not only in its fize, which is always inferior, but by its bill, which is longer, by the yellow colour of its legs, and above all, the habit of changing its refidence. It has been feen in Brie, and in other places, paffing in numerous flocks, and purfuing its journey without halting. A game-keeper in the neighbourhood of Montbard faw laft March (1770), a flight of one hundred and fifty or two hundred, which feemed to turn afide and fufpend their progrefs at the noife of the call, but were entirely gone the next day. This fimple fact, which is undoubted, points out the analogy

and

and the difference between this fpecies and the common fort. Their being drawn by the call fhewed their affinity; their rapid flight through a country equally fuited to the Gray and Red Partridges, which both refide in it the whole year, denotes another inftinct, and confequently another organization, and at leaft another family.

We muft not confound this Damafcus or Syrian Partridge with the *Syroperdix* of Ælian, found in the vicinity of Antioch; whofe plumage was black, the bill fulvous, the flefh firmer and better flavoured, and the difpofition more favage than that of other Partridges: for the colours we fee do not correfpond, and Ælian does not tell us that this *Syroperdix* is a bird of paffage. He adds, as a fingular circumftance, that it fwallows ftones; which however is very common in the granivorous tribes. Scaliger mentions a remarkable fact, to which he was witnefs, and which bears fome relation to the prefent; it is that in Gafcony, where the land is very fandy, their flefh was mixed with minute particles, which was very difagreeable. [A]

[A] Specific character of the Damafcus Partridge, *Tetrao Damafcenus* :—" It has a naked faffron fpot under its eyes, its tail " is ferruginous, its breaft brown, its feet yellow."

The MOUNTAIN PARTRIDGE.

Le Perdrix de Montagne, Buff.
Tetrao Montanus, Gmel.
Perdix Montana, Briff.

I make this Partridge a diftinct fpecies, fince it refembles neither the Gray nor the Red fort. It would be difficult to decide to which of thefe kinds we ought to refer it ; for if, on the one hand, it be certain they fometimes breed with Gray Partridges; on the other, their ordinary refidence is on mountains ; and the red colour of their bill and legs, alfo fhews a clofe relation to the Gray Partridge, and I am ftrongly inclined to fufpect that they fometimes even confort with thefe. I am therefore perfuaded that it conftitutes the intermediate fpecies between thefe extremes ; it is nearly the fize of the Gray Partridge, and has twenty quills in the tail.

[A] Specific character of the Mountain Partridge, *Tetrao Montanus :* — " Its feet and bill are red, its throat tawny-yellowifh."

The RED PARTRIDGES.

The GREEK PARTRIDGE.

La Bartavelle, ou *Perdrix Grecque*, Buff.
Tetrao Rufus, Linn. and Gmel.
Perdix Græca, Briff.
Pernice, Zinn.

Whatever the ancients have faid on the fubject of Partridges, we muft refer to the Red
kinds, and efpecially to the *Bartavelle*. Ariftotle
was undoubtedly beft acquainted with the Greek
Partridge, nor is it probable that he knew any but
the Red Partridges; fince thefe are the only Partridges that are found in Greece, or in the iflands
of the Mediterranean *, and in all probability in
the part of Afia conquered by Alexander, fituated
in nearly the fame latitude with Greece and the
Mediterranean †, and which was probably the
fource of Ariftotle's principal information. With
refpect to the fucceeding naturalifts, fuch as
Pliny, Athenæus, &c. we plainly fee that though
they were acquainted with other Partridges in
Italy befides the Red, they were contented with
barely copying what Ariftotle had faid. It is

* Belon.

† It appears that only the Red Partridge was known to the
Jews, fince they reprefent it as an inhabitant of the mountains :
" The king of Ifrael is come out to feek a flea, *as one would hunt*
" *a Partridge on the mountains.*" 1 Samuel, chap. 26. ver. 20.

indeed

indeed true that the Grecian philofopher admitted a difference in the cries of the Partridges *; but we cannot thence infer any real diftinction of fpecies: for this diverfity often refults from the age or fex, has place even in the fame individual, and may be the effect of fome local caufe or of the influence of climate; which the ancients themfelves admitted, fince Athenæus afferts that the Partridges which paffed from Attica into Bœotia were known to change their cry †. Befides, Theophraftus, who alfo remarks fome varieties in the notes of the Partridges, according to the countries which they inhabit, certainly fuppofes them not to be all of different fpecies; for he defcribes the different voices in his treatife " *On the various Notes of Birds of the fame kind.*"

On examining the accounts which the ancients have given refpecting this bird, I difcover many accurate facts and obfervations disfigured by a heap of exaggerations and fables; on which fome moderns ‡ have fhewn their pleafantry, though it required no great talents to ridicule. I fhall endeavour to trace the origin of thefe tales from the nature and inftincts of the Partridge.

Ariftotle relates that this bird is fond of rolling in the duft, has a craw, a gizzard, and very fmall *cæca* §; that it lives fifteen years and

---

* Some Κακαβιζυσι, others Τριζυσι.   † Gefner.
‡ Willughby.   § *Hift. Anim.* lib. ii. cap. ult. and lib. vi. 4.

more*; that, like all other birds of a laborious
flight, it builds no neft, but lays its eggs on the
open ground, on a little herbage or leaves ftrewed
carelefsly †, though in a fpot of good afpect, and
fheltered from the attacks of rapacious birds;
that in this fpecies, which is of a very amorous
difpofition, the males fight obftinately with each
other in love feafon, and that at that time their
tefticles are diftinctly feen, though in winter ‡
they are hardly vifible; that the females can lay
eggs without any intercourfe with the male §;
that both fexes copulate by opening the mouth
and darting out the tongue ‖; that their hatch
commonly confifts of twelve or fifteen eggs;
that fometimes they cannot retain their eggs,
but drop them wherever they happen to be ¶.
But after mentioning thefe facts, which are in-
conteftible, and which are confirmed by the obfervations of the moderns, Ariftotle adds many
circumftances where the truth is difguifed, and
which muft be analifed, in order to extract what
is valuable from the mixture.

He fays, 1. That the female Partridges lay
moft of their eggs in a concealed fpot, to fave

---

* Lib. ix. 7. Gefner has inadvertently put twenty-five years
in his verfion, which error has been copied by Aldrovandus.
Athenæus makes Ariftotle fay that the female lives longer than
the male, as ufual in birds.

† Lib. vi. 1. ‡ Lib. iii. 1. § Id. ibid.

‖ Lib. v. 5. Avicenna has thence been led to fay, that the
Partridges work up their paffion by the clofeft kiffes and careffes,
like the pigeons; but this is a miftake.

¶ Lib. ix. 8.

them

them from the male, who feeks to deftroy them, as impeding his pleafures *. This is reckoned ridiculous by Willughby; but I am inclined to think that he has been too hafty in paffing judgment, for if we diftinguifh between the fact obferved and the intention implied, the affertion of Ariftotle is literally true, and is nothing more than that the Partridge, like all the other females of the feathered race, is induftrious to conceal her neft; left the males, efpecially the fupernumerary ones, feeking to copulate in the time of incubation, difturb the function by the gratification of their appetites. For this reafon it has always been recommended to deftroy the fuperfluous males, as one of the moft efficacious means of advancing the multiplication of the breed, not only of Partridges, but of other birds in the favage ftate.

Ariftotle adds, 2dly, That the female Partridge divides her eggs into two hatches, one of which fhe entrufts to the male, until the young are educated †. This is abfolutely incompatible with the propenfity to break the eggs, which he fuppofes to be implanted in the male. But if we would reconcile Ariftotle with himfelf and with truth, we may fay, that as the female does not lay all her eggs in the fame fpot; fince fhe cannot retain them, but allows them fometimes to drop wherever fhe happens to fit; and as the male feems in this fpecies, or at leaft in fome

---

* Id. ibid.   † *Hift. Anim.* lib. vi. 8.

families

families of this species, as in the gray sort, to share
in the education of the young, it is not impro-
bable but he also participates in the drudgery of
incubation, and may cover a part of the eggs
which were not under the hen.

Aristotle says, 3dly, That the males tread one
another, and even their young as soon as they
are able to walk *; and this assertion has been
considered as fabulous and absurd. I have however more than once had occasion to mention
undoubted instances of this perversion of instinct; and, among the Partridges, nature is so
purient, that the male cannot hear the cry of the
female without shedding *semen*; and so intoxicated is he with pleasure in the season of love,
that though commonly extremely shy, he ven-
tures then to alight on the bird-catcher. How
much more therefore would their ardour be ex-
alted in the warm climate of Greece, and when
they had long been denied the company of their
mates, as during the time of incubation?

Aristotle asserts, 4thly, That female Partridges
conceive and produce eggs, when they happen to
be fanned by wind from the males, or when
these fly over them, or even when they hear
their notes †. These words of the Grecian philosopher have given occasion to much ridicule;
as if a current of air impregnated with the fœcun-
dating particles of the male, or the mere vibra-

* Lib. ix. 8.   † Lib. v. 5.

tion

tion impreffed by his voice, were really fufficient
to impregnate the female. All that is infinuated
is, that in fuch circumftances the natural fire of
their conftitution blazes with new force; and
for the fame reafon, weltering in the duft forwards the laying *.

From thefe facts it is eafy to conceive, that
the Hen Partridge, though ftrongly fet on covering, will fometimes prefer the indulgence of
appetite to the tedious duty of incubation. It
may even happen that, when fhe perceives her
mate wavering in his fidelity, and about to yield
to the allurements of other females, fhe will
offer the embrace, to fecure the domeftic harmony, and provide for the profperity of the expected progeny †.

Ælian has faid, that the males fight always with
greater obftinacy in prefence of the females; becaufe, he adds, when thus circumftanced, they
will rather die, than fhew cowardice, or appear
after being vanquifhed ‡. We muft here too
diftinguifh between the fact and the intention.
It is certain that the fight of the females adds
fuel to their quarrels ; not however becaufe it

---

* Ariftotle adds a fact which evinces their falacious temperament ;
" they alfo lay *zephyrian* or addle eggs, if the genital parts be
" ftroked by the finger."

† Often the female rifes from her neft when fhe perceives her
mate attending to a fauntering female, and throwing herfelf into
his embrace, fatiates his appetite. Arift. lib. ix. 18.—So that luft
overcomes even the attachment to their brood. Lib. x. 33.

‡ Hift. Anim. lib. iv. 1.

piques

piques their honour, but becaufe it inflames their appetite for enjoyment.

It is thus by diftinguifhing between actions and intentions, and rejecting crude fuppofitions which disfigure important facts, that we can often extract the truth from relations of animals, which have fo much been deformed by the fictions of man, and the folly of afcribing to all other beings his own character, his own feelings, and his own apprehenfions.

As the *Bartavelles* poffefs many properties in common with the Gray Partridges, we need only remark the chief differences. Belon, who had vifited their native climate, tells us, that they are double the fize of our Partridges: that they are very plentiful, and even more common than any other bird in Greece, in the Archipelago iflands, and particularly on the coafts of the ifland of Crete (now Candia): that they call in the love feafon, uttering a found like that of the word *chacabis ;* whence the Latins have undoubtedly formed the term *cacabare* to denote that cry ; and the fame analogy might perhaps have had fome fhare in the conftruction of the *cubeth, cubata, cubeji,* &c. the names of the Red Partridge in the eaftern languages.

Belon relates alfo, that the *Bartavelles* generally lodge in rocks, but that they come down into the plains to build their neft, in order that their young may procure food with eafe: that they lay from eight to fixteen eggs, of the fize of fmall

hen

hen eggs, white and sprinkled with reddish
points, and the yolk of which cannot be made
hard. Lastly, what he imagines shews the Greek
Partridge to be of a different kind from our Red
Partridge, is this, that in Italy, where both are
known, they have each a distinct name, the
Grecian sort being called *Cothurno*, and the other
Perdice; as if the vulgar who bestow names could
not be mistaken, or even apply two different
denominations to the same species. He conjectures also, and not without probability, that
it is this large Partridge, which, according to
Aristotle, crosses with the ordinary hen and
breeds prolific hybrids; a circumstance which, as
the Greek philosopher remarks, rarely happens,
and never but between the most salacious kinds[*].
It bears still another analogy to the common
hen, that it sits on other eggs when robbed of
its own: —This observation is very ancient, for
it occurs in scripture [†].

Aristotle observes, that the male Partridges
sing or cry chiefly in the love season, when they
fight with each other, or even before they begin
to quarrel [‡]. The ardour which they have for
the female is then converted into mutual rage.

---

[*] *De Generatione Animalium,* lib. ii. 4.

[†] " As the Partridge gathereth the young which she has not
" brought forth ; so he that getteth riches, and not by right, shall
" leave them in the middle of his days, and in the end be a
" fool."

[‡] Lib. iv. 9.

Hence

Hence thofe contefts, and thofe fcreams, that intoxication, and that heedlefs fecurity which precipitates them not only into fnares, but into the
very hands of the fowler *.

Fowlers have profited by their ardent headlong difpofition to draw them into the fnare: a
female is prefented to their burning appetites, or
a male has been expofed to provoke their im-
prudent rage †. The males have even been
trained to fight by way of entertainment, and
thefe birds, commonly fo peaceable and fo timid,
have contended with obftinate fury, and the
combat has been inflamed by the fight of the
females ‡. This cuftom is ftill very common in
the ifland of Cyprus §; and we have already obferved that the Emperor Alexander Severus
took great delight in this fort of battles. [A]

* Lib. ix. 8.   † Lib. iv. 1.
‡ Ælian *de Nat. Anim.* lib. iv. 1.
§ *Hift. de Chypre,* par Francois Stephano Lufignano.

[A] Specific character of the Greek Partridge, *Tetrao Rufus :* —
" Its legs and bill are blood-coloured; its throat is white, en-
" circled with a black belt, dotted with white."

M

The

The EUROPEAN RED PARTRIDGE.

Tetrao Rufus, Var. Linn. and Gmel.
Perdix Rufa, Var. Lath. and Ind.
Perdix Rubra, Briff.
The Red-legged Partridge, Ray, and Will.
The Guernfey Partridge, Lath. Syn.

This is of an intermediate fize between the *Bartavelle* and the Gray Partridge. It is not fo common as the latter, nor does every climate fuit it. It is found in moft of the temperate and mountainous countries of Europe, Afia, and Africa: it is rare in the Netherlands *, and in many parts of Bohemia and Germany, where the attempts to breed it have proved unfuccefsful, though the pheafant thrives there †. It is never feen in England ‡, nor in certain iflets near Lemnos §; yet a fingle pair, being carried into the little ifland of Anapha (now Nanfio), multiplied to fuch a degree, that the colonifts were almoft refolved to abandon their fettlements ‖. This abode is fo congenial to their nature, that, even at this day, the inhabitants are obliged, about Eafter holidays, to deftroy their eggs by thoufands; left the Partridges, which might be hatched, fhould totally ruin the crops. And

* Aldrovandus.   † Idem.   ‡ Ray and Edwards.
§ Anton. Liberalis *apud Aldrov*.   ‖ Athenæus.

thefe

THE GUERNSEY PARTRIDGE

thefe eggs prepared with different fauces ferve
feveral days to fubfift the iflanders *.

The Red Partridges fettle in mountains which
abound with heath and bufhes, and fometimes
in the fame mountains which are inhabited by
the game improperly termed *White Partridges;*
yet they prefer the lower and more hofpitable
tracts †. In winter, they confine the range of
their excurfions, and lodge under the fhelter of
rocks with a fouthern afpect. During the reft of
the year they continue in the bufhes, and the
fportfmen can hardly drive them from their retreats. I am well informed that they can, better
than the common Partridge, fupport the rigours
of winter, and are much more eafily caught by
gins or fnares. They refort every fpring in
nearly the fame numbers to their favourite haunts.
They feed on grain, herbs, flugs, caterpillars,
ants' eggs, and other infects; but their flefh is
often tainted with the fmell of their aliments.
Ælian relates that the Partridge of Cyrrha, a
maritime town in Phocis, had a difagreeable
tafte, becaufe of their living on garlick.

They fly heavily and laborioufly, like the gray
fort; and without feeing them, we may eafily
diftinguifh them by the noife merely which
they make with their wings when they are
flufhed. When they are furprifed on the mountains, they feek fhelter among the precipices, and

* Tournefort.   † Stumpfius *apud Gefn.*

when

when they are diflodged, they regain the fummit.
In the plains, they fhoot fwiftly forward. When
they are hotly purfued, they fly into the woods,
and perch upon the trees, and fometimes even
burrow in the ground, which the Gray Partridge never does.

The Red Partridges are diftinguifhed from
the Gray alfo by their natural habits and difpofitions; they are not fo focial : they form themfelves indeed into coveys, but the union is not
fo complete or harmonious. Though hatched
and bred together, the Red Partridges keep
apart from each other : they do not fpring at
the fame inftant, they do not fly in the fame direction, and they do not call each other with the
fame eagernefs, except in the love feafon, and
then even each pair forms a feparate union.
Laftly, When their paffion is gratified, and the
female begins coolly to cover her eggs, the male
abandons to her the charge of raifing the family.
In this refpect, our Red Partridges feem to differ
from thofe of Egypt; fince the priefts chofe as
the emblem of domeftic harmony, a pair of Partridges, a male and a female, occupied each apart
with its hatch *.

A confequence of the favage difpofition
of the Red Partridge, is that they are more
difficult to breed in parks like the pheafant,
though the method is nearly the fame. It requires more pains and attention to habituate

* Aldrovandus.

them

them to their captivity: nor are they ever completely reconciled, fince the young Partridges languifh in their prifon, and though every expedient be tried to fweeten their condition, would pine away or fall into fome diforder, if not fet at liberty as foon as their feathers begin to fhade their heads.

Thefe facts, which have been communicated to me by M. Le Roi, feem to contradict what is related of the Partridges of Afia *, and fome iflands in the Archipelago †, and even thofe of Provence ‡, where numerous flocks have been feen that obeyed the voice of their conductor with wonderful docility.   Porphyry mentions a tame Partridge brought from Carthage, which ran to call his mafter, fawned on him, and expreffed its

---

* " In the country round Trebizond, I faw a man leading above
" four thoufand Partridges. He marched on the ground, while the
" Partridges followed him in the air, till he reached a certain camp
" three days journey from Trebizond.   When he flept, all the
" Partridges alighted to repofe around him, and he could take as
" many as he pleafed of their number."
Odoricus de Foro-Julü, apud Gefn.

† " There are people on the coaft of Veffa and Elata (in the
" ifle of Scios), who raife Partridges with care.   They lead them
" to feed in the fields, like flocks of fheep : each family entrufts
" its Partridges to the common keeper, who brings them back in
" the evening; and he calls them together by means of a whiftle,
" even in the day-time."
Tournefort's Voyage to the Levant.

‡ I have feen a man in Provence who led flocks of Partridges
into the fields, and affembled them whenever he chofe ; he took
them with his hand, put them into his bofom, and then difmiffed
them. Id. ibid.

fondnefs

fondnefs by certain articulations which feemed to flow from fentiment, and were entirely different from its ordinary notes. Mundella and Gefner raifed fome themfelves that grew very familiar. It appears from feveral paffages in ancient authors, that they had even acquired the art of teaching them to fing, or at leaft to improve their natural notes fo much as to give a pleafing fort of warble *.

But all this may be reconciled, by faying that this bird has not fo great averfion to man as abhorrence of flavery: that he has difcovered the way to tame and fubdue the moft favage animal, that is, one the moft enamoured of liberty; and that the method is to treat it according to its inftinct, and to indulge it with as much freedom as poffible. In this light, the fociety of the tame Partridge with the perfon who directs its will, is the moft engaging, and the moft noble; founded not on its wants, its interefts, or on a ftupid ferenity of temper, but bound by fympathy, choice, and a mutual affection. The Partridge contracts a liking for man, and fubmits to his inclinations, only when he allows it to leave him at pleafure, and impofes no reftraint beyond what fociety requires. In a word, when he attempts to reduce it to domeftic flavery, its generous nature revolts at the appearance of force; the lofs of liberty preys upon its vitals, and ex-

* Athenæus.—Plutarch.—Ælian.

tinguifhes

tinguifhes the moft powerful inftincts, thofe of love and of felf-prefervation. Sometimes, in the paroxyfms of its rage, it dafhes its head againft the cage and expires. It difcovers an invincible repugnance to propagation: and if fometimes, ftimulated by the ardour of temper, and the warmth of the feafon, it copulates in confinement, its embraces are too feeble to perpetuate a race of flaves. [A]

[A] The Red Partridge is not found in England; but in France it is the moft common of the genus.

The WHITE RED PARTRIDGE.

In the fpecies of the Red Partridge, as in the Gray, the plumage is fometimes white; which change of colour is the accidental effect of fome particular caufe. Nor is this whitenefs intimate; the colour of the head is not altered, and the bill and legs remain red: and as they commonly breed with the Red Partridges, we have reafon to conclude that they belong to the fame fpecies.

The FRANCOLIN.

Tetrao Francolinus, Linn. and Gmel.
Perdix Francolinus, Lath. Ind.
Tetrao Orientalis, Haſſelq.
The Francoline Partridge, Lath. Syn.

This name, too, has been beſtowed on very different birds. We have already ſeen it applied to the *Attagas*; and from a paſſage of Geſner, the bird known at Venice by the name of *Francolin*, appears to be a kind of Hazel Grous*.

The Neapolitan Francolin is larger than a common hen; and indeed the length of its legs, bill, and neck, will not allow us to regard it as either an Hazel Grous or a Francolin †.

All that we know of the Francolin of Ferrara is, that it has red feet, and lives on fiſh ‡. The bird of Spitzbergen, which has been called Francolin, receives alſo the appellation of the *Beach Runner*, becauſe it never ſtrays far from the ſhore, where it picks up gray worms and ſhrimps for its ſubſiſtence: it is no larger than a lark §. The Francolin which Olina figures and deſcribes, is the one of which I am to treat. That of Edwards differs from it in ſome reſpects,

---

* It is the ſame with the hazel-hen of the Germans, which I diſcovered beyond all doubt from a figure of the Venetian Francolin ſent to me by the learned phyſician Aloyſius Mundella.
GESNER.

† Ibid.   ‡ Ibid.   § *Voyages de Prevôt*, tome xv.

and

and appears to be exactly the same bird with the Francolin of Tournefort, which also resembles that of Ferrara, since it is found on the sea-coast, and in marshy situations.

Ours seems to differ from these three last, and even from that of Brisson, not only in the colour of its plumage, and even of its bill, but by the size and form of its tail, which is longer in Brisson's figure, more spread in ours, and hanging in those of Edwards and Olina. But notwithstanding this, I believe that the Francolin of Olina, that of Tournefort, that of Edwards, that of Brisson, and my own, are all of the same species; since they have many common properties, and their small differences are not sufficient to constitute different races, but may be referred to the age, the sex, the climate, and other local or accidental circumstances.

The Francolin is undoubtedly, in many respects, like the Partridges, and for this reason, Olina, Linnæus, and Brisson, have ranged it with them. For my own part, I am convinced, from a close examination and comparison of these two birds, that they ought to be separated: for the Francolin differs from the Partridge not only in the colours of its plumage, its general shape, the figure of its tail, and its cry; but is distinguished also by a spur on each leg [*]; whereas the male Partridge has only a callous tubercule.

[*] That of Olina had none; but it was probably a female.

The Francolin is alfo much lefs diffufed than the Partridge; it can hardly fubfift but in the warm climates. Spain, Italy, and Sicily, are almoft the only countries of Europe where it is found; it inhabits alfo Rhodes*, the Ifle of Cyprus †. Samos ‡, Barbary, efpecially in the vicinity of Tunis §, Egypt, the coafts of Afia ‖, and Bengal ¶. In all thefe places, both Francolins and Partridges occur; but they have each their appropriated name, and form diftinct fpecies.

As thefe birds are very rare in Europe, and their flefh is excellent food, the killing them has been forbidden in many countries under fevere penalties; and hence, it is faid, they derive the name *Francolin;* becaufe they enjoy a fort of freedom under the protection of thefe prohibitions.

Little more can be faid of this bird than what the figure fuggefts: its plumage is very beautiful; it has a confpicuous collar of an orange colour. It is rather larger than the Common Partridge; the female is fmaller than the male; the colours of its plumage fainter and lefs variegated.

Thefe birds feed on grain; and they may be bred in aviaries, though care muft be taken to give each a fmall feparate crib, where it may

* Olina.   † Tournefort.   ‡ Edwards.
§ Olina.   ‖ Tournefort.   ¶ Edwards.

fquat

fquat and conceal itfelf, and to ftrew fand and a little gravel on the floor.

Their cry is a kind of loud whiftle, audible at a great diftance *.

Francolins live much in the fame manner as Partridges †; their flefh is exquifite, and fometimes preferred to that of Partridges or Pheafants.

Linnæus takes the Damafcus Partridge of Willughby for the Francolin ‡. But we may obferve, 1ft, That this Damafcus Partridge is rather Belon's, who firft took notice of it, and whom Willughby only copied. 2dly, This bird differs from the Francolin, both in the fize, which is inferior to that of the Common Partridge, according to Belon; and in its plumage, as will eafily be perceived by comparing the figures; and befides, its legs are feathered, which prevented Belon from claffing it with the plovers. Linnæus fhould alfo have admitted the Francolin of Tournefort as the fame with that of Olina, which Willughby mentions. Laftly, the Swedifh naturalift is miftaken in fixing exclufively on the Eaft as the climate of the Francolin; for, as I have already obferved, it is found in Sicily, Italy, Spain, and Barbary, and in many other countries to which the epithet of oriental cannot be applied.

* Olina.   † Ibid.
‡ Tenth edition of the *Syftema Naturæ.*

Ariftotle

Ariſtotle ranges the *Attagen*, which Belon conceives to be the Francolin, among the pulverulent and granivorous birds. Belon makes him alſo ſay, that it lays a great number of eggs, though no mention of this ſort is made in the place quoted ; but it is the neceſſary conſequence of Ariſtotle's theory with regard to pulverulent granivorous birds. Belon relates, on the authority of the ancients, that the Francolin was common in the plain of Marathon, being fond of marſhy ſituations ; which agrees very well with Tournefort's obſervations reſpecting the Francolins at Samos. [A]

[A] Specific character of the *Tetrao Francolinus* :—" Its belly " and throat are black, its tail wedge-ſhaped."

DOUBLE SPUR.

Le Bis-Ergot, Buff.
Tetrao Bicalcaratus, Linn.
Perdix Bicalcaratus, Lath. Ind.
Perdix Senegalenſis, Briſſ.
The Senegal Partridge, Lath. Syn.

The firſt ſpecies which ſeems to approach to the Francolin, is the bird to which, in the *Planches Enluminees*, we have given the name of *Senegal Partridge*. It has, on each foot, two ſpurs, or rather tubercles, of hard, callous

fleſh ;

flefh; and as it is a diftinct fpecies, we may call it *Double Spur*, on account of that fingular cha-racter.   I place it next the Francolins, as being more related to them than to the Partridge; by its fize, by the length of its bill and wings, and by its fpurs.   [A]

[A] Specific character :—" Its legs are double fpurred ; its " eye-brows black."

The B A R E - N E C K E D

A N D

AFRICAN   RED   PARTRIDGE.

Tetrao Nudicollis, Gmel.
Perdix Nudicollis, Lath.

This bird, which we have feen alive at Paris at the houfe of the late Marquis de Montmirail, had the lower part of the neck and throat di-vefted of feathers, and merely covered with a red fkin ; the reft of the plumage was much lefs variegated and lefs pleafant than that of the Francolin.   It refembles that fpecies by its red legs and the fpreading fhape of its tail ; and is related to the preceding fpecies, by the double fpur on each leg.

c c 3   The

The want of facts prevents me from inquiring into the analogy between thefe two fpecies in refpect to their difpofitions and habits. M. Aublet affures me that it is a bird which never perches.

The AFRICAN RED PARTRIDGE has more red than our fpecies, becaufe of a broad fpot of that colour under its throat; but the reft of its plumage is much inferior. It is diftinguifhed from the two preceding by two very obvious characters; its fpurs are long and pointed, and its tail more expanded than is common in Partridges. We have not obfervations fufficient to enable us to judge whether they differ alfo in their modes of living. [A]

[A] Specific character :—" Its legs are double fpurred and " rufous ; its throat naked and rufous."

FOREIGN BIRDS,

THAT ARE RELATED TO THE PARTRIDGE.

I.

THE RED PARTRIDGE OF BARBARY, Pl. LXX. of Edwards*, feems to be a different kind from the European Red Partridge, and is fmaller than the Gray. Its bill, its orbits, and its feet, are red, as in the Greek Partridge; but the fcapular feathers are of a fine blue, edged with brown-red; and round the neck is a fort of collar formed by white fpots, fcattered on a brown ground, which, joined to its diminutive fize, diftinguifhes this fpecies from the two breeds of Red Partridges common in Europe.

* *Tetrao Rufus*, Var. 3. Gmel.; *Perdix Rubra Barbarica*, Briff.

II. The

II.

The ROCK PARTRIDGE,

OR

GAMBRA PARTRIDGE.

Tetrao Petrofus, Gmel.
Perdix Petrofa, Lath. Ind.
The Rufous-breafted Partridge, Lath. Syn.

This bird takes its name from its favourite haunts; it delights, as do the Red Partridges, in rocks and precipices; its predominant colour is a dull brown, and it is marked on the breaft with a fpot like the colour of Spanifh tobacco. It alfo refembles the Red Partridge in the colour of its legs, its bill, and its orbits; it is fmaller than our fort, and cocks its tail when it runs, but is, like them, very fleet; its fhape, on the whole, is the fame, and its flefh excellent *. [A]

* *Journal de Stibbs.* PREVÔST.

[A] Specific chara&ter of the *Tetrao Petrofus :*—" Its bill and " legs are red, its body dufky, with a ferruginous fpot on its " breaft."

III. The

III.

The PEARLED CHINESE PARTRIDGE.

Tetrao Perlatus, Gmel.
Perdix Sinenfis, Briff.

This Partridge, known only by Briffon's defcription, feems to be peculiar to the eaftern
extremity of the Antient Continent. It is fomewhat larger than the Red Partridge of Europe,
but has its fhape, the figure of its tail, the fhortnefs of its wings, and the fame general appearance. Of the Common Red Partridge it has the
white throat; and of the African fort, the long,
pointed fpurs; but it has not, like that bird, the
red bill and legs: thefe are here of a ruft-colour,
and the bill is blackifh, as well as the nails. The
ground of the plumage is dufky, enlivened on
the breaft and fides by a number of fmall round
light-coloured fpots: from this property I have
applied the name of *Pearled Partridge*. It has,
befides, four remarkable bars, which rife from
the bottom of the bill and ftretch over the fides
of the head; thefe bars are alternately of a
bright and deep colour. [A]

[A] Specific character of the *Tetrao Perlatus*:—" Its legs and
" eye-brows are rufous, its bill blackifh, its throat white, its
" body dufky and variegated."

IV. The

IV.

The NEW ENGLAND PARTRIDGE.

Tetrao Marilandus, Linn. and Gmel.
Perdix Marilanda, Lath. Ind.
Perdix Novæ Angliæ, Briff.
The American Partridge, Du Pratz.
The Maryland Partridge, Penn. and Lath.

I refer this American bird, and the following, to the Partridges; not that I imagine them to be real Partridges, but only the reprefentatives: for of the birds in the New World they approach the neareft to the Partridge, though it is impoffible that this fpecies could wing its courfe over the immenfe oceans which feparate the continents.

This bird is fmaller than the common Gray Partridge; its iris is yellow, its bill black, its throat white, and two bars of the fame colour ftretch from the bafe of the bill to the back of the head, paffing over the eyes. It has alfo fome white fpots on the top of the neck; the under-fide of the body is yellowifh, ftriped with black, and the upper fide of a brown bordering on rufous, nearly as in the Red Partridge, and checquered with black; its tail is fhort, as in the other Partridges. It is found not only in New England, but in Jamaica, though thefe two climates differ widely.

Albin

Albin fed one a confiderable time with wheat and hemp-feed. [A]

[A] Specific character of the *Tetrao Marilandus* :—" Its eye-
" brows are white, its neck dotted with black and white." The
American Partridges are about one-half larger than the Englifh
Quails, and are plump and well-tafted. They are frequent in
North America, as high as Canada. They lay from twenty to
twenty-five eggs, and breed about the beginning of May ; their
numerous covies make a loud noife when fprung. The cock
perches on a fence, and emits his double note while the hen is
fitting. Thefe birds have been introduced into Jamaica, where
they are naturalized, and faid to breed twice a-year.

M

The Q U A I L *.

La Caille, Buff.
Tetrao-Coturnix, Linn. and Gmel.
Perdix Coturnix, Lath. Ind.
Coturnix, All the Naturalifts.

THEOPHRASTUS difcerned fuch intimate relation between the Partridge and the Quail,
that he beftowed on the latter the name of
Dwarf Partridge. The fame appearance of
analogy muft have led the Portuguefe to call
the Partridge *Codornix;* and the Italians apply
the term *Coturnice* to the *Bartavelle,* or Greek
Partridge. Thefe birds, indeed, refemble each
other confiderably : they are both pulverulent;
they have fhort wings and tail, and run very
fwiftly; their bill is like that of the gallinaceous
tribe; their plumage is gray, fpeckled with
brown, and fometimes entircly white †. Befides, they feed, copulate, build their neft, hatch
their eggs, and raife their young nearly in the
fame way; both are of a falacious difpofition,
and the males quarrel much with each other.

---

* In Greek, "Ορτιξ; in Latin, *Coturnix*; in Italian, *Quaglia*;
in Spanifh, *Cuadervix*; in German, *Wachtel.* Frifch afferts,
that in the time of Charlemagne it was called *Quacara*; fome
others have termed it *Currelius,* on account of its fwift running.

† Ariftotle, *de Coloribus,* chap. vi.

But

THE QUAIL .

But how numerous foever be the points of refemblance, they are balanced by an almoft
equal number of difparities, which make the
Quails a fpecies entirely diftinct from the Partridges. For, 1. The Quails are univerfally
fmaller than the Partridges, comparing correfponding breeds. 2. They have not the bare
fpace between the eyes, as in the Partridges,
nor the figure of the horfe-fhoe impreffed on the
breaft of the males, nor have true Quails the
bill and legs ever red. 3. The eggs are fmaller,
and of an entirely different colour. 4. Their
notes alfo are quite different, though they love
in nearly the fame feafon; but the Partridges
intimate their rage before they fight, while the
Quails fcream only in the midft of their quarrels *. 5. The flefh of the latter is of a different tafte, and much more loaded with fat.
6. The period of their life is much fhorter.
7. They are lefs cunning than the Partridge,
and much more eafily enfnared, efpecially
when young and unexperienced. Their difpofitions are not fo gentle; it is extremely rare
to fee them tamed, and though confined from
their infancy, they can hardly be trained to
obey the voice. They are not of fuch a focial
temper; for they feldom form themfelves into
coveys, except when their wants unite the feeble
family to their mother, or fome common and
powerful caufe urges at once the whole fpecies

* Ariftotle, *Hift. Aaim.* llb. viii. 12.

to affemble together, and traverfe the extent of the ocean, holding their courfe to the fame diftant land. But this forced affociation fubfifts no longer than neceffity requires; and after they have alighted, and find in their adopted country that they can live at will, their union diffolves. The appetite of love is the only tie that binds them together, and even this connection is unftable and momentary; for though the male feeks the female with the greateft ardour, he difcovers no choice or predilection; the matches are formed haftily, and as quickly broken. As foon as paffion has fpent its force, the male treats his mate with indifference and cruelty, and abandons her to the labour and care of raifing the family. The young are hardly grown up when they feparate, or if they are kept together, they fight obftinately with e_ h other; their quarrels are terminated only by their mutual deftruction *.

The propenfity of the Quails to migrate at certain feafons, is one of their moft powerful inftincts.

The caufe of this defire muft be very general, for it acts not only on the whole fpecies, but alfo on individuals kept in confinement and debarred from communication with their kind. Some young Quails, bred in cages from their

---

* The ancients were well acquainted with this fact, for they faid that obftinate quarrelfome children were like Quails in a cage.
ARISTOPHANES.

earlieft

earlieft infancy, which had never enjoyed liberty, and therefore could not feel its lofs, were yet obferved, for the fpace of four years, to fhew a degree of reftleffnefs, and to flutter with unufual agitations, regularly at the feafon of migration, which returns twice annually, in April and in September. This uneafinefs lafted thirty days each time, and began conftantly an hour before fun-rife. The prifoners moved backward and forward from one end of the cage to the other, and darted againft the net which covered it, and often with fuch violence, that they dropped down ftunned by the blow. They paffed the night in thefe fruitlefs ftruggles, and the following day they appeared dejeſted, exhaufted, and torpid. We know alfo that, in the ftate of liberty, Quails fleep the greater part of the day: and if to this we add, that they are feldom obferved to arrive in the day-time, we may infer, that they perform their journies by night *, and that the difpofition to migrate is innate: whether that avoiding the extremes of heat and cold, they remove to the north in fummer, and advance to the fouth in winter; or what is more probable, that they direſt their courfe to thofe countries where the harveft is making, and thus change their abode to pro-cure the proper fubfiftence for themfelves and for their young.

* Belon and Pliny exprefs the fame opinion.

This

This laft reafon, I fay, is the moft probable; for, on the one hand, it is proved by experience, that Quails can well fupport cold; fince Horrebow informs us that they have been found in Iceland; and they have been kept for years together in a room without fire, and having even a northern afpect, and yet feemed not in the leaft affected by the fevereft winters. On the other hand, it appears, that one circumftance which determines them to abide in a country, is the plenty of herbage; for it is well known by fportfmen, that when the fpring is dry, and confequently grafs fcarce, there are few Quails the reft of the year. Befides, the fpur of actual want is a more powerful caufe, is more confiftent with the limited inftinct of thefe animals, and implies lefs forefight, which philofophers have rather been too liberal in beftowing on brutes. When they cannot procure fubfiftence in one country, it is very natural to fuppofe that they will feek it in another. This fcarcity of food intimates to them their fituation, and roufes all their faculties into action. They leave the exhaufted tract, rife into the regions of air, and pufh forward to difcover countries which may prefent them with abundance. Habit joining itfelf to the inftinct common to all animals, but moft remarkable in the winged tribes, of fcenting their food from a

diftance,

diftance, gives birth to a difpofition which is as it were innate; it is not therefore furprifing that the fame Quails fhould return annually to the fame haunts. But we can hardly fuppofe with Ariftotle, that it is from an attentive ob-fervation of the feafons, and that they change their climate twice a-year, like the ancient kings of Perfia. It will be ftill more difficult to ad-mit, with Catefby and Belon, and fome others, that when they fhift their refidence, they wing their courfe without interruption to the Anti-podes, there to find exactly the fame latitude, and to enjoy the fame temperature; which would imply fcientific knowledge, or rather error, to which brute inftinct is much lefs fub-ject than cultivated reafon.

But whatever fpeculations we may form with refpect to the migration of the Quails, certain it is, that when they enjoy their natural liberty they depart and return at ftated times. They left Greece according to Ariftotle, in the month *Boedromion* *, which comprehended the end of Auguft and the beginning of September. In Silefia, they arrive in the month of May, and depart about the end of Auguft †. Our fportfmen reckon on their return to France about the tenth or twelfth of May. Aloyfius Mundella fays, that they begin to appear in the neighbourhood of Venice about the middle of

* Hift. Anim. lib. viii. 12.　　† Schwenckfeld.

　　April.

April. Olina fixes the time of their arrival in the *Campagna di Roma* in the beginning of April. But almoft all agree that they depart after the firft autumnal froft, which fpoils the grafs and deftroys the infects; and fince the colds of May do not drive them back to the fouth, we are furnifhed with another proof that warmth is not what they feek, but that their real object is food, and of which the fpring frofts cannot deprive them. We muft not however confider thefe terms as invariably fixed. They will vary within certain limits in different countries according to the nature of the climate; and even in the fame region, they will be affected by the latenefs or earlinefs of the feafon, which will advance or retard the harveft, and will promote or check the multiplication of the infects which fupport the Quails.

Both the ancients and the moderns have been bufy in forming theories with regard to the migration of the Quail and other birds of paffage; fome have heightened it by the addition of the marvellous; while others, ftruck with the difficulty of conceiving that fo fmall a bird, and which flies tardily and laborioufly, could perform diftant journies, have hefitated to admit the fact, and have had recourfe to hypothefes ftill more extraordinary to account for their regular difappearance at certain ftated feafons. None of the ancients ever entertained indeed the flighteft doubt on the fubject: and yet they

well

5

well knew that the Quail is inactive, flies little and with reluctance*; and that though extremely ardent in its paſſions, it employs not its wings to tranſport itſelf to the invitation of the female, but often travels more than a quarter of a league through the cloſeſt herbage to meet her, and never riſes into the air except when cloſely purſued by dogs or hunters : with all theſe circumſtances, the ancients were acquainted; but they never dreamt that, on the approach of the cold ſeaſon, theſe birds crept into holes to remain in a dormant ſtate during the winter, like the dormice, the hedge-hogs, the marmots, the bats, &c. This abſurdity was reſerved for ſome moderns, who certainly did not know that the internal heat of animals ſubject to become torpid, being generally inferior to that of other quadrupeds, and ſtill more to that of birds, requires a conſtant acceſſion of warmth from the air, as I have ſhewn in another work: when this ſource fails, the vital action is ſuſpended; and, if they were expoſed to a greater cold, they would in a ſhort time periſh. But this certainly is not applicable to the Quails, which are generally eſteemed of a hotter conſtitution than the other birds; ſo that in France it has given riſe to a proverb †; and in China, it is cuſtomary to carry this bird alive in one's hands to keep them

---

* Ariſtotle, *Hiſt. Anim.* lib. viii. 8.

† It is a common ſaying, *Chaud comme un Caille,* warm as a Quail.

warm.

warm *. Besides, I have discovered from observations continued for several years, that they
never grow torpid, though shut during the whole
winter in rooms without a fire, and exposed to
the north, as I have formerly mentioned : and
many persons of the most undoubted veracity,
who had similar opportunities, have assured me
of the same fact. But, if the Quails neither
conceal themselves nor remain torpid through
the winter, and always disappear in that season,
we may certainly conclude that they migrate into
other countries.—And this fact is demonstrat-
ed by a great number of other authorities.

Belon, happening in autumn to be on board a
vessel, in his passage from Rhodes to Alexandria,
saw Quails flying from the north towards the
south. Many of them were caught by the
mariners, and grains of wheat were found very
entire in their craw. The preceding spring, the
same observer saw in sailing from the island of
Zante to the Morea, a great number of them in
motion from south to north; and he affirms
that, in Europe as well as Asia, Quails are generally migrating birds.

The Commander Godeheu constantly saw
them passing Malta in the month of May, aided
by certain winds, and again in September in their
eturn †. Many sportsmen have assured me,

* Osborn's Travels.

† Memoires de Mathematique and de Physique, presentes a
l'Academie Royal des Sciences par divers Savans, &c. tome iii. 91.

that

that they have often, in the fine nights during spring, heard them arrive, and could easily distinguish their cry, though at a great height in the air. Add to this, that they are no where so plenty as on the French coasts, opposite to Africa or Asia, and in the interjacent islands. Almost all those of the Archipelago and even the shelves, are, according to Tournefort, covered with them, in certain times of the year ; and more than one of those islands has been named *Ortygia* *. In the age of Varro, it was remarked that at the seasons of the arrival and departure of the Quails, immense flocks were seen in the islets of Pontia, Pandataria, and others scattered along the southern coast of Italy †, and which they probably chose for a station to recruit. About the beginning of autumn, such great numbers were caught in the island of Caprea, in the entrance of the Gulph of Naples, that the bishop of the island drew his principal revenue from the profits of the game, and was for that reason called *The Bishop of Quails*. Many too are caught in the neighbourhood of Pesaro, which is situated on the Adriatic Gulf, about the end of autumn, the season of their arrival ‡. Lastly, such amazing numbers drop

---

* From Ορτυξ, which signifies a Quail. The two Delos were, according to Phanodemus in Athenæus, termed *Ortygiæ*: so also another little island opposite to Syraeuse, and even the city of Ephesus, according to Stephanus of Byzantium and Eustathius.

† De Re Rustica, lib. iii. 5.

‡ Aloysius Mundella, *apud Gesnerum*.

on

on the weſtern coaſts of the kingdom of Naples,
in the vicinity of *Nettuno*, that in the extent of
four or five miles, ſometimes a hundred thouſand
are taken in a day, and are ſold for fifteen *jules*
the hundred (leſs than ſeven ſhillings), to a ſort
of brokers who carry them to Rome, where
they are much leſs common *. Clouds of them
alſo alight in the ſpring on the coaſts of Provence,
eſpecially on the lands belonging to the biſhop
of Frejus, which border on the ſea; they are exhauſted, it is ſaid, with the fatigue of their journey, that for the firſt days they may be caught
with the hand.

But it will ſtill be replied, how can a bird ſo
ſmall, ſo weak, whoſe flight is ſo ſlow and laborious, how can it, though urged by hunger, tra-
verſe the great extent of ſea? I may admit that
many iſlands are ſcattered at intervals in their
paſſage, on which they may halt to recruit their
vigour: ſuch as Minorca, Corſica, Sardinia, Sicily,
Malta, Rhodes, and other iſles in the Archipelago.
But ſtill I conceive that it would be impoſſible for
them to perform the journey without aſſiſtance.
Ariſtotle was well convinced that this was neceſſary, and was even acquainted with the kind
of aid which the Quails moſt commonly received;
and if he was miſtaken, it is only in deſcribing
the manner. " When the north wind blows,

---

* Geſner and Aldrovandus. This capture is ſo lucrative, that
land near the place is extravagantly high priced.

" the

" the Quails, (fays he,) perform their journey
" with eafe ; but when the fouth wind prevails, as
" it oppreffes them with the load of its vapours,
" they make a painful progrefs, and declare
" their labour and exertion, by the cries which
" they utter in their flight *." In fact, it is the
wind, I conceive, which affifts the Quails in
their paffage ; not indeed the north wind alone,
but a favourable wind; nor does the fouth
impede their progrefs, but fo far as it is
contrary to the direction of their motion : and
this muft take place in all countries where the
Quails perform their journies acrofs the feas †.

M. Godeheu has well remarked, that in
the fpring, the Quails never alight on Malta,
except when they are carried by a north-weft
wind, which hinders them from gaining Provence ; and that in their return, they are wafted
to that ifland by the fouth wind, which oppofes
their defcent on the Barbary fhore ‡. We know
alfo, that the Author of nature employed that
mean, as the moft conformable to the general
laws which he had eftablifhed, to fhower the
immenfe multitudes of Quails upon the Ifraelites
in the defert § ; and this wind, which came from
the fouth-weft, fwept over Egypt, Ethiopia, and
the coafts of the Red Sea, and in a word, the

* Lib. viii. 12.   † Pliny expreffes the fame idea, lib. x. 23.
t Memoires prefentés a l'Acad. &c. tome iii. 92.
§ Pfalm lxxvii.

countries

countries where thefe birds were moft abundant *.

Sailors whom I have confulted on this fubject inform me, that when the Quails are furprifed in their paffage by a contrary wind, they
alight on the neareft veffels, as Pliny has remarked †, and often fall into the fea, and are
then obferved to float and ftruggle on the waves,
with one wing raifed in the air to catch the
gale. Hence fome naturalifts have taken occafion to fay, that when they embark on their
voyage, they furnifh themfelves with a little
ftick, with which they relieve themfelves at intervals from the fatigue of flying, refting upon
it as upon a raft, and riding on the rolling bil-
lows ‡. It has even been fuppofed that each
carries in his bill three fmall ftones, to ballaft
them, according to Pliny §, againft the violence
of the wind; or, according to Oppian ‖, to difcover ¶, by dropping them one after another,
when they have croffed the fea. This is nothing
more than bits of gravel which the Quails, like
other granivorous birds, fwallow with their food.
In general, fuch a degree of defign, fagacity,
and difcernment, is afcribed to them, as would

* The Gulf of Arabia abounds very much with Quails.
JOSEPHUS, lib. iii. 1.
† " They arrive not without danger to navigators who have ap-
" proached near the land ; for they alight on the fails, and this al-
" ways at night, and fink the veffels." PLINY Hift. Nat. lib. x. 23.
‡ See Aldrovandus, tome ii. 116.   § Lib. x. 23.
‖ In Ixeut.   ¶ Pliny, lib. x. 23, and Solinus, cap. xvii.
give

give room to doubt if thofe who are fo liberal in beftowing thefe qualities, really poffefs a large fhare themfelves. They have remarked that other migratory birds, as the Land Rail, accompany the Quails, and that the Falcon was fure to catch fome one on its arrival; hence they have concluded that the Quails choofe out from another fpecies a guide or chief, which they call *The King of the Quails (ortygrometra)*; for as the foremoft of the body falls a victim to the hawk, the Quails fhrewdly contrive to caft the danger upon the fated individual of a foreign race.

But the Quails do not all migrate; there are a few which, being unable to follow the reft, remain behind; either having received a wound in their wings, or, being the product of the fecond hatch, they are too young and feeble to perform the journey. Thefe ftragglers feek to find a proper fituation in the country where they are obliged to abide *. In France the number of thefe is very inconfiderable; but the Author of the Britifh Zoology affures us, that in England a part only of the Quails are obferved to quit the ifland entirely, while thofe which remain fhift their quarters, removing from the interior counties to the fea coaft, and particularly to the hundreds of Effex, where they continue through the winter: if the froft or fnow drive

* Aldrovandus, lib, viii. 12.

them

them from the ftubble fields or marfhes, they retreat to the beach, and fubfift upon the marine plants, which they can pick up between high and low water mark. The fame Author fub-joins, that the time of their appearance in Eſſex correſponds exactly to that of their leaving the inland country. It is likewiſe faid, that a great number of them remain in Spain and in the fouth of Italy, when the winter is not fo fevere as entirely to deftroy the feeds and infects that ferve for their food.

With refpect to fuch as venture to crofs the feas, thoſe only perform a fortunate voyage that are affifted by a fair wind; and if in the pre-ceding feafon it has feldom blown from a favour-able quarter, much fewer arrive in the countries where they fpend their fummer. And, in ge-neral, we may judge with tolerable accuracy of the place whence they have come, by the direc-tion of the breeze which wafts them along.

As foon as the Quails arrive, they fet about lay-ing. They do not pair, as I have already re-marked; and if the number of males, as I am informed, exceeds much that of the females, monogamy would have been inconvenient. Fide-lity, confidence, perfonal attachment, qualities fo defirable in the individual, would have been hurtful in the fpecies. The multitude of males reduced to a ftate of celibacy, would difturb the marriages that are formed, and render them un-prolific.

prolific. But in the prefent cafe, the unbounded liberty of union blunts and extinguifhes the jealoufy and the rivalfhip of their loves. The male has been obferved to repeat a dozen times a-day his embraces with feveral females in-difcriminately *; and while nature tolerates this kind of libertinifm, fhe provides for the mul-tiplication of the fpecies. Each female drops fifteen or twenty eggs into a neft, which fhe hollows in the ground with her claws, lines with grafs and leaves, and conceals as much as pof-fible from the piercing eye of the hawks. Thefe eggs are greyifh, fpeckled with brown. She fits about three weeks. The ardor of the males effectually fecundates them, and they are feldom found addle.

The Authors of the Britifh Zoology fay, that in England the Quails feldom lay more than fix or feven eggs. If this fact be uniform, we may conclude that they are lefs prolific in that ifland than in France, Italy, &c.; and it remains to be inquired whether this diminution of the genial powers ought to be attributed to the cold, or fome other quality of the climate.

The young Quails are able to run almoft as foon as they leave the fhell, like the young Par-tridges; but they are in many refpects more hardy, fince in the ftate of liberty they quit their mother much earlier, and even venture to de-

* Aldrovandus and Schwenckfeld,

pend

pend on themfelves for fubfiftence eight days
after they are hatched. This circumftance has
made fome perfons fuppofe that the Quail lays
twice a year *; but I much doubt it, except
they have been difturbed in their firft hatch.
It is not even affirmed that they begin another after they have arrived in Africa in the
month of September, though this would be much
more probable, fince on account of their regular
migrations they are unacquainted with autumn
and winter, and the year to them confifts of two
fprings and two fummers; and they change
their climate, it might feem, to enjoy and perpetuate the ever verdant feafon of love and propagation.

Certain it is, that they drop their feathers twice
a year, in the end of winter and on the approach
of autumn. Each moulting lafts a month; and
as foon as their plumage is reftored, they wing
their paffage, if at liberty, into other climes; and
if they are kept in confinement, they difcover a
reftleffnefs and ftruggle to burft from their
prifon at the ftated periods of migration.

When the young are four months old, they
are able to accompany their parents in their
diftant flight.

The female differs from the male in being
fomewhat larger, according to Aldrovandus

* Aldrovandus afferts, that the Quails begin to lay the fame
year they are hatched, in the month of Auguft, and have ten
eggs.

(others

(others reckon it equal and others fmaller), in having a white breaft fprinkled with black fpots almoft round ; while in the male it is reddifh, without any intermixture of other colours : its bill is alfo black, as well as its throat, and a few hairs that grow round the bafe of the upper mandible * : its tefticles have alfo been remarked to be very large in proportion to the fize of the body ; but this obfervation has undoubtedly been made in the feafon of their amours, when the tefticles of all birds acquire a confiderable increafe of dimenfions.

The male and female have each two cries ; a louder and a fainter. The male makes a found like *ouan, ouan, ouan, ouan ;* he never gives the fonorous call, except when parted from the females ; nor when confined, though he has a female companion. The female has a well-known cry, which invites the male ; and though it is very weak and audible only a fhort way, the males flock to it from the diftance of half a league : it has alfo a flender quivering note *cri, cri.* The male is more ardent than the female, as he runs to her amorous murmurs with fuch precipitation and inconfideratenefs, as to alight to her upon the hand of the bird-catcher †.

* Aldrovandus.—Some naturalifts have taken the male for the female : I have on this occafion followed the opinion of fportfmen, efpecially thofe of obfervation.

† Ariftotle, lib. viii. 12.

The

The Quail, like the Partridge and many other animals, never multiplies its kind, except when it enjoys the liberty of nature. In vain have the forlorn prisoners been furnished with the materials for constructing their nests; the tenderest concerns are stifled in their breast, and their eggs are allowed to drop unheeded.

Many foolish stories have been told with respect to the generation of Quails. It has been said that, like the Partridges, they are impregnated by the wind; this means no more than that they sometimes lay without the male influence*. It has been alleged that they are bred from the tunnies, which the troubled ocean sometimes casts on the shores of Lybia; that they appear at first like worms, then assume the form of flies, and gradually growing larger, they become grasshoppers, and at last Quails †: that is, the vulgar seeing the Quails searching among the carcases of the tunnies rejected by the waves for some insects that are hatched in them, and having some vague idea of the metamorphoses which insects undergo, fancied that a grasshopper could be transformed into a Quail, as a worm is changed into a fly. Lastly, it has been said, that the male copulates with the female toad ‡; a story which has not even the slightest appearance of probability.

* Ibidem.　　　　† Gesner.
‡ Phanodemus *apud Gesuerum.*

The

The Quails feed on wheat, millet, hemp-feed, green herbage, infects, and all forts of feed, even that of the hellebore; which gave the antients an antipathy to its flesh, augmented still further by the reflection that thefe were the only animals befides man that are fubject to the epilepfy *. But experience has deftroyed this prejudice.

In Holland, where thefe birds are frequent, efpecially along the coafts, it is ufual to call the berries of Bryony Quail-berries; which fhews that they prefer that fort of food †.

It would appear that they can fubfift without drink; for fportfmen have aſſured me, that they are never obferved to feek water; and others, that they have fed them a whole year on dry grain without any drink, though they frequently drank when it was in their power. To with-draw every kind of drink, is even the only way to cure them when they *caſt out their water;* that is, when they are attacked by a certain dif-temper, in which they have always a drop at the point of their bill.

Some have imagined, that they always trouble the water before they drink, and they have not failed to afcribe it to envy; for naturalifts are never fatisfied till they aſſign to brutes the mo-tives of action. They inhabit the fields, the pafture grounds, and the vineyards; but feldom

* Pliny, lib. x. 23.　　† Hadrian. Jun. *Nomenclat.*

refort

refort to the woods, and never perch upon trees.
They grow much fatter than Partridges.   What
is fuppofed to contribute to this, is their re-
maining ftill during the greateft part of the heat
of the day; then they conceal themfelves in the
talleft grafs, and fometimes continue in the fame
fpot for the fpace of four hours, lying on their
fide, with their legs extended ; and fo much are
they overcome with the drowfy indolence, that
a dog muft abfolutely run upon them before
they are flufhed.

It is faid that they live only four or five years,
and Olina attributes the fhortnefs of the term to
their difpofition to corpulency.   Artemidorus
imputes it to their unhappy quarrelfome temper ;
and fuch is really their character, for they have
been made to fight in public to entertain the
rabble.   Solon even directed that fuch combats
fhould be exhibited to the youth, with the view
to inflame their courage.   And this fpecies of
gymnaftic exercife, which would appear to us fo
puerile, muft have been held in high eftimation
by the Romans, and confidered as an affair of
the ftate, fince we are told that Auguftus punifh-
ed a prefect of Egypt with death, for buying
and bringing to table one of thefe birds that had
acquired celebrity by its victories.   Even at
prefent this fort of amufement is common in
fome cities of Italy. They give two Quails high
feeding; and then place them oppofite to each

other,

other, at the ends of a long table, and throw between them a few grains of millet feed (for they
need a ground of quarrel). At firſt they ſhew a
threatening afpect, and then ruſh on like light-
ning, ſtrike with their bills, erecting the head
and riſing upon their ſpurs, and fight till one
yields the field of battle*. Formerly, theſe
combats were performed between a Quail and a
man: the Quail was put into a large box, and
ſet in the middle of a circle traced on the floor;
the man ſtruck it on the head with one finger,
or plucked ſome feathers from it: if the Quail,
in defending itſelf, did not paſs the limits of the
circle, its maſter gained the wager; but if in
its fury it tranſgreſſed the bounds, its worthy
antagoniſt was declared victor; and ſuch Quails
as often won the prize ſold very dear †. It may
be remarked that theſe birds, as well as the Partridges and others, never fight but with their
own ſpecies; which implies jealouſy, rather than
courage, or even violence of temper.

Since the Quail is accuſtomed to migrate, and
travels to immenſe diſtances by the aid of the
wind, it is eaſy to conceive that it muſt be ſpread
through a wide extent. It is found at the Cape
of Good Hope, and through the whole inhabited part of Africa ‡; in Spain, Italy §, France,
Switzerland ||, the Netherlands ¶, Germany **,

* Aldrovandus. † Julius Pollux *de Ludis*, lib. ix.
‡ Kolben, and Joſephus, lib. iii. 1. Comeſtor, &c.
§ Aldrovandus. || Stumpfius.
¶ Aldrovandus. ** Friſch.

England,

England *, Scotland †, Sweden ‡, and as far as Iceland §; and eaftwards, in Poland ‖, Ruffia ¶, Tartary **, as far as China ††. It is even poffible that it could migrate into America; fince it every year penetrates near the polar circles, where the two continents approach; and, in fact, it occurs in the Malouine iflands, as we fhall afterwards take notice. In general, it is more common along the coafts than in the interior country.

The Quail is therefore an univerfal inhabitant, and is every where efteemed excellent game. Aldrovandus tells us, that the fat is fometimes melted by itfelf, and kept for fauce.

The female, or a call imitative of her cry, is made ufe of to draw the males into the fnares. It is even faid, that a mirror having a noofe placed before it is fufficient; the bird, miftaking its image for another of the fpecies, rufhes towards it. The Chinefe catch them as they fly with flender nets, which they ufe very dextroufly ‡‡. In general all the forms of gins that are ufed for other birds, fucceed with the Quails,

* Britifh Zoology. † Sibbald. ‡ Linnæus *Fauna Suecica*.
§ Horrebow. ‖ Rzaczynfki. ¶ Cramer and Rzaczynfki.
** Gerbillon, "Travels performed into Tartary, in the fuite "or by the order of the Emperor of China." *Hift. Gen. des Voyages*, tome vii. p. 465. and 505.
†† Edward's Gleanings, vol. i. The Chinefe, fays he, have alfo our common Quail, as evidently appears from their paintings, in which it is depicted from nature.
‡‡ Gemelli Carreri.

and

and efpecially the males, which are lefs fufpicious, more ardent, and which may be led at pleafure by imitating the cry of the female.

This ardor of the Quails has occafioned the quality to be afcribed to their eggs, fat, &c. of reftoring a relaxed frame and roufing the genial powers *. It has been faid that the prefence alone of one of thefe birds in a bed-chamber, gave thofe who flept there love dreams †.—We need only quote thefe ftories, as they refute themfelves. [A]

* " The eggs of the Quail rubbed on the tefticles procure " pleafure, and if fwallowed they ftimulate luft." KIRANIDES.

† Frifch.

[A] Specific character of the Quail, *Tetrao Coturnix :*—" Its " body is fpotted with gray, its eye-brows white, the margin of " its tail-quills, with a crefcent, ferruginous." The Quail occurs in every part of Great Britain, but is not frequent.

The CHROKIEL, OR THE GREAT POLISH QUAIL.

Tetrao Coturnix, Var. 1. Gmel.
Coturnix Major, Briff.

Our knowledge of this Quail is drawn from the Jefuit Rzaczynfki, a Polifh author, who merits the more attention on this fubject, as he

defcribes

defcribes a bird which is a native of his own
country. In its fhape, and even its habits, it
exactly refembles the Common Quail, and differs only by its fize; and for this reafon I confider it as merely a variety.

Jobfon fays, that the Quails of Gambra are
as large as Wood-cocks*; and if the climate
were not widely different, I fhould confider
them as the fame with the Polifh fort.

The WHITE QUAIL.

Tetrao Coturnix, Var. 2. Gmel.

Ariftotle is the only naturalift who mentions
this Quail†, which muft be viewed as a variety;
juft as the grayifh-white and white-red partridges are varieties of thefe two fpecies of the
partridge, and the white lark a variety of the
common lark, &c.

Martin Cramer fpeaks of Quails‡ with greenifh legs; is this a variety of the fpecies, or merely
adventitious in the individual?

* Purchas's Collection of Voyages, vol. ii.
† *De Colonibus,* cap. 6.   ‡ *De Polonia,* lib. i. 474.

The

The QUAIL of the MALOUINE ISLANDS.

Tetrao Falklandicus, Gmel.

We may confider this bird as a variety of the common fort which is diffufed through Africa and Europe, or at leaft a proximate fpecies; the only difference being, that its plumage is of a deeper brown, and its bill fomewhat ftronger.

But what oppofes this idea, is the immenfe expanfe of ocean which feparates the two continents towards the fouth : our Quails muft have performed an aftonifhing voyage, if we fuppofe they held their courfe from the north of Europe to the Straits of Magellan. I will not therefore decide whether this Quail is the fame fpecies with ours, or only a branch from the fame ftem, or if not, rather a breed peculiar to the Malouine Iflands. [A]

[A] Specific character of the *Tetrao Falklandicus* :—" It is va-
" riegated with dufky curved ftreaks and fpots, below, white ; its
" bill lead coloured, its feet dufky, its temples fpotted with
" white."

The

The RUFF, or CHINESE QUAIL.

Tetrao Chinenſis, Linn. and Gmel.
Coturnix Philippenſis, Briſſ.

This bird is figured in the *Planches Enluminées* by the name of the *Quail of the Philippines,* becauſe it was ſent from theſe iſlands to the Royal Cabinet. But it is alſo found in China, and I have called it the *Ruff,* on account of a ſort of white ruff under its neck, which is the more remarkable, as its plumage is of a brown verging upon black. Edwards gives a figure of the male, Pl. CCXLVII.: it differs from the female in our *Planches Enluminées,* in being ſomewhat larger, though ſtill not bigger than a lark; its aſpect is alſo more marked, the colour of its plumage more lively and variegated, and its feet ſtronger.—The ſubject, which is deſcribed by Edwards, was brought alive from Nankin to England.

Theſe little Quails have this character in common with the ordinary ſorts, that they fight obſtinately with each other, particularly the males: and the Chineſe lay conſiderable bets, as cuſtomary in England on game cocks. We cannot therefore heſitate to admit that they are of the ſame genus with our Quails, though probably of a different ſpecies.

The

The TURNIX, or MADAGASCAR QUAIL.

Tetrao Striatus, Gmel.

We have given this Quail the name *Turnix*, contracted for *Coturnix*, to diftinguifh it from the ordinary kind, from which it differs in many refpects. For, 1ft, it is fmaller; 2dly, its plumage is different both in the colours and their diftribution; and, 3dly, it has three fore-toes on each foot, like the buftards, and none behind. [A]

[A] Specific character of the *Tetrao Striatus :*—" Its legs " tawny; its eye-brows white; its bill, its throat, the lower part " of its breaft, and its belly, black, with white drops."

―――――――――

The N O I S Y Q U A I L.

Réveil-Matin *, ou *La Caille de Java,* Buff.
Tetrao Sufcitator, Gmel.
Coturnix Javenfis, Briff.

This bird is not much larger than our Quail, refembles it exactly in the colours of its plumage, and pipes at intervals; but it is diftinguifhed by many notable differences.

* i. e. Morning Waker.—See Bontius's " Natural and Medical " Hiftory of the Eaft-Indies."

1. Its

1. Its note is very deep, and very ftrong, and pretty much like the fort of lowing of the bittern, ingulphing its bill in the marfhes *.

2. Its difpofition is fo gentle that it can be tamed to the fame degree as our domeftic fowls.

3. It is remarkably affected by cold; it ceafes to pipe, and its active powers are fufpended, in the abfence of the fun. As foon as he has defcended into the weft, it retires into fome hole, and fpends the night enveloped in its wings; but when the ftar of day again beams upon the earth, it rifes from its lethargy, and celebrates his return with joyous notes, that awaken the whole houfe †. Alfo, when kept in a cage, if it has not the fun conftantly, or if the cage is not covered with a coat of fand upon linen cloth to retain the heat, it will pine away and foon die.

4. Its inftinct is different; for, according to Bontius's account, it is very focial, and goes in companies. Bontius adds, that he found it in the forefts on the ifland of Java; but our Quails live folitary, and are never found in the woods.

5. Its bill is fomewhat longer.

---

* The Hollanders call this lowing *Pittoor*, according to Bontius.

† Bontius fays, that he kept one in a cage for the exprefs purpofe of rouzing him in the morning: in fact their firft calls announce always the rifing of the fun.

This

This fpecies has however one point of analogy to our Quail, and to many others ; to wit,
the males fight each other with exceffive rancour,
and defift not till one is killed.——But this circumftance is not a fufficient foundation for ar-
ranging it with thefe, and I have therefore beftowed on it a diftinct name. [A]

[A] Specific character of the *Tetrao Sufcitator* :—" It is varie-
" gated with yellowifh, rufous, black, and gray ; its bill longer."

OTHER BIRDS

WHICH ARE RELATED TO THE PARTRIDGES
AND THE QUAILS.

I.

The COLINS.

THE Colins are Mexican birds, which have
rather been mentioned than defcribed by
Fernandez *; and thofe who have copied that
author on this fubject have committed fome
miftakes, which it will be proper to correct.

First, Nieremberg †, who profeffes to take his
accounts entirely from others, and who in this
place borrows from Fernandez, takes no notice
of the *Cacacolin* of chap. cxxxiv. though that
bird is of the fame fpecies with the Colins.

Secondly, Fernandez fpeaks of two *Acolins*,
or Water-Quails, in chap. x. and cxxxi.; Nieremberg mentions the former, and very improperly, after the Colins; fince it is a water-bird, as
well as the one of chap. cxxxi. which he totally
omits.

---

* *Hiftoria Avium Novæ Hifpaniæ*, cap. xxiv, xxv. xxxix. lxxxv.
and cxxxiv.

† *Joan Eufeb. Nirembergi Hiftoria Naturæ maxime Peregrinæ*,
lib. x. cap. lxxii. p. 232.

Thirdly,

Thirdly, He takes notice of the *Occolin* of chap. lxxxv. of Fernandez, which is a Mexican Partridge, and confequently nearly related to the Colins, which are alfo Partridges, as we fhall fee.

Fourthly, Ray ftill copying Nieremberg, on the fubject of the *Coyolcozque* varies the expreffion, and in my opinion alters the meaning of the paffage; for Nieremberg fays, that this Coyolcozque is like the Quails fo called by us *Spaniards*, (which are certainly the Colins,) and concludes with telling that this is a fpecies of the Spanifh Partridge. But Ray makes him fay that it is like the European Quails, and fuppreffes the words *eft enim fpecies perdicis Hifpanicæ* * ; yet thefe laft words are effential, and contain the real notion of Fernandez with regard to the fpecies to which thefe birds muft be referred; fince, in chap. xxxix. which is occupied entirely on the Colins, he fays that the Spaniards call them *Quails*, becaufe they refemble the European Quails, though they certainly belong to the genus of Partridges. It is true, that he repeats in the fame chapter, that all the Colins are referred to the Quails; but, in fpite of this confufion, it is eafy to fee that when the author beftows on the Colins the name of *Quails*, he fpeaks after the vulgar, who are guided in applying epithets by the general appearances, and that his more ac-

* i. e. *For it is a fpecies of Spanifh Partridge.*

I 2   curate

curate opinion was, that they are species of the Partridge. I should therefore have had reason, from the authority of Fernandez, the only obferver who has had an opportunity of viewing these birds, to place the Colins next the Partridges; but I have rather chosen to yield as much as possible to the common opinion, which is not altogether groundless, of ranging them after the Quails, as being related to both these kinds of birds.

According to Fernandez, the Colins are very common in New Spain; their music resembles much that of our Quails; their flesh is excellent, and proper even for sick people when kept some days. They feed on grain, and are commonly kept in a cage; which would make one believe that they are different from our Quails, and even our Partridges.—We shall in the following articles take notice of their several kinds.

II.

The ZONECOLIN*.

Tetrao Criftatus, Linn. and Gmel.
Coturnix Mexicana Criftata, Briff.
The Crefted Quail, Lath.

This word, shortened for the Mexican *Quanht-zonecolin,* denotes a bird of a moderate size, whose

* Fernandez, chap. xxxix.

plumage

plumage is of a dufky colour; it is diftinguifhed by its cry, which, though rather plaintive, is agreeable, and by the creft which decorates its head.

Fernandez mentions, in the fame chapter, another Colin of the fame plumage, but not fo large and without the creft; this is perhaps the female of the preceding, from which it is diftinguifhed only by accidental charaƈters, that are liable to vary in the different fexes. [A]

[A] Specific charaƈter of the *Tetrao Criftatus :—*" Its pendulous " creft and its throat are fulvous."

III.

The GREAT COLIN*.

Tetrao Novæ Hifpaniæ, Gmel.
Coturnix Major Mexicana, Briff.
The Mexican Quail, Lath.

This is the largeft of all the Colins: Fernandez does not give us its name; he only fays that its predominant colour is fulvous, that its head is variegated with white and black, and that there is alfo white on the back and on the tips of the wings, which muft make a fine contraft with the black colour of its legs and bill.

* Fernandez, chap. xxxix.

IV. The

IV.

The   C A C O L I N*.

This bird is called the *Cacolin* by Fernandez, and is, according to him, a fpecies of Quail, that is of the Colin, of the fame fize, fhape, and even cry; feeding on the fame fubftances, and having its plumage painted with almoft the fame colours with thofe of the Mexican Quails. Neither Nieremberg, Ray, nor Briffon, takes any notice of it.

V.

The   C O Y O L C O S.

Tetrao Coyoleos, Gmel.
Coturnix Mexicana, Briff.
The Leffer Mexican Quail, Lath.

I have foftened the Mexican word *Coyolcozque* into this name. This bird, in its cry, its fize, its habits, its manner of living and of flying, refembles the other Colins, but differs from them in its plumage. Fulvous mixed with white, is

* " Species of what is called the Quail." *Fernandez,* chap. cxxxiv.

the

the prevailing colour of the upper fide of the
body, and fulvous alone that of the under fide
and of the legs : the top of the head is black and
white, and two bars of the fame colour defcend
from the eyes upon the neck : it inhabits the
cultivated fields.—Such is what Fernandez relates, and Briffon muft have read the account
with little attention, or rather copied Ray, when
he tells us that the Coyolcos is like our Quail in
its cry, flight, &c.; while Fernandez exprefsly
fays, that it is analogous to the Quails, fo called
by the vulgar, that is to the Colins, and is really
a fpecies of the Partridges *.   [A]

* " It is a fpecies of the Spanifh Partridge."   *Hift. Anim.*
Novæ Hifpaniæ.

[A] Specific charaƈter of the *Tetrao Coyolcos :* " Its feet are
" fulvous, its top and its neck are ftriped with black and white ;
" its body is fulvous above, variegated with white."

VI.

The C O L E N I C U L I.

Tetrao Mexicanus, Linn. and Gmel.
Coturnix Ludoviciana, Briff.
Attagen Americanus, Frifch.
The Louifiana Quail, Lath.

Frifch gives (Pl. CXIII.) the figure of a bird,
which he calls *The Small Hen of the forefts of
America,* and which, according to him, refembles
the

the Wood Grous in its bill, legs, and general form;
its legs however are not feathered, nor are its toes
edged with indentings, nor its eyes decorated
with red orbits, as we may fee from the figure.
Briffon, who conceives this bird to be the fame
with the *Colenicuiltic* of Fernandez, has ranged
it among the Quails, by the name of *Louifiana
Quail*, and gives a figure of it. But comparing
the figures or the defcriptions of Briffon, Frifch,
and Fernandez, I find greater differences than
could occur in the fame bird; for not to mention
the colours of the plumage, fo difficult to paint in
defcription, and ftill lefs the attitude, which is but
too arbitrary, I obferve that the bill and the legs
are large and yellowifh, according to Frifch;
red and moderate fized, according to Briffon;
and that the legs are blue, according to Fernandez.

But if I attend to the different lights in which
naturalifts have viewed it, the embarraffment will
be increafed; for Frifch fancied that it was a
Hen of the Wood, Briffon a Quail, and Fernandez a Partridge. That this was the opinion of
the laft manifeftly appears, for though he fays, in
the beginning of chap. xxv. that it is a Quail,
he evidently conforms to the common language;
fince he concludes the chapter with faying, that
the *Colenicuiltic*, in its bulk, in its cry, in its
habits, and in every other particular, is analogous
to the bird of chap. xxiv; but that bird is the
Coyolcozque, a kind of Colin; and Fernandez,

as

as we have already feen, ranks the Colins among the Partridges.

I would not infift on this matter, were it not to avoid as much as poffible the great inconvenience attending on nomenclature. Each author, fond of building a fyftem, is not fatisfied till he affign to every object, however anomalous, its place; and thus, according to the different views that arife, the fame animal may be claffed with *genera* widely diftinct.—Such is the prefent cafe.

To return—The Colenicui is of the bulk of our Quail, according to Briffon; but its wings feem to be longer; its body is brown above, and dirty-gray and black beneath; it has a white throat, and a fort of white eye-brows. [A]

[A] Specific character of the *Tetrao Mexicanus:*—"Its legs and "bill are blood-coloured, the line on its eye-brows white."

VII.

The OCOCOLIN, or MOUNTAIN PARTRIDGE of MEXICO.

This fpecies, which Seba took for the crefted Roller of Mexico, is ftill farther removed from the Quail, and even the Partridge, than the preceding. It is much larger, and its flefh is not in-

ferior to that of the Quail, though much inferior to that of the Partridge. The Ococolin
refembles fomewhat the Red Partridge, in the
colour of its plumage, of its bill, and of its feet;
its body has a mixture of brown, light gray, and
fulvous ; the lower-part of its wings is of an afh-
colour, the upper-part is mottled with dull white,
and fulvous fpots, as likewife the head and neck.
It thrives beft in countries that are temperate and
rather chilly, and cannot fubfift or propagate in
the hot climates.—Fernandez fpeaks alfo of another Ococolin, but which is a bird entirely of
a different kind *.

* " Ococolin, a kind of Wood-pecker with a long fharp bill."
It lives in the forefts of Telzcocan, where it breeds : it does not
chirp.

The

The PIGEON DOMESTICA.

Columba, Linn. and Gmel.

IT was eafy to domefticate the heavy and inactive birds, fuch as the common hen, the
turkey, and the peacock; but to tame thofe which
are nimble and fhoot on rapid wings, required
attention and art. A low hut, rudely conftructed
on a confined fpot, is fufficient for lodging and
raifing our poultry; to induce the Pigeons to
fettle, we muft erect a lofty building, well covered without and fitted up with numerous cells.
They really are not domeftics, like dogs or
horfes; or prifoners like the fowls; they are
rather voluntary captives, tranfient guefts, who
continue to refide in the dwelling affigned them,
only becaufe they like it, and are pleafed with a
fituation which affords them abundance of food,
and all the conveniencies and comforts of life.
On the flighteft difappointment or difguft, they
abandon their manfion, and difperfe; and fome
of them even will always prefer the mouldering
holes of ancient walls to the neateft apartments
in Pigeon-houfes; others take their abode in
the clefts and hollows of trees; others feem to
fly the habitations of men, and cannot be pre-
vailed to enter their precincts; others again never

F F 2

roam

roam from human dwellings, but muſt be fed near their volery, to which they are inflexibly attached. Theſe various and even oppoſite habits ſhew, that under the Pigeon are included many different ſpecies. This opinion is confirmed by the modern nomenclators, who, beſides a great number of varieties, reckon five ſpecies of Pigeons, without including the Ring-dove and Turtle. We ſhall remove theſe two laſt ſpecies from thoſe of the Pigeon, and conſider each ſeparately.

The five ſpecies of Pigeons noticed by our nomenclators are, 1. The Domeſtic Pigeon; 2. The Roman Pigeon, which includes ſixteen varieties; 3. The Brown Pigeon; 4. The Rock Pigeon, with one variety; 5. The Wild Pigeon; but theſe five ſpecies are in my opinion the ſame. My reaſon is this. The Domeſtic Pigeon and the Roman Pigeon, with all their varieties, though differing in ſize and colours, are certainly the ſame ſpecies; ſince they breed together, and their progeny are capable of procreating. We cannot conſider the great and little Domeſtic Pigeons as two different ſpecies; we can only ſay that they are different branches of the ſame kind, the one of which has been reduced to a more perfect domeſtication than the other. In the ſame manner, the Brown Pigeon, the Rock Pigeon, and the Wild Pigeon, are three nominal ſpecies which may be compriſed in one, which is the Brown Pigeon, and of which the Rock Pigeon and the Wild Pigeon are only minute varieties;

ſince

since the nomenclators themselves admit that these three are nearly of the same size, that they migrate, perch, and have all the same instincts, differing only in their shades of colour.

Thus the five nominal species are comprised under two; viz. the Brown Pigeon and the Common Pigeon; and in these no real difference exists, except that the first is wild, and the second domestic. I consider the Brown Pigeon as the parent of all the rest, and from which they differ more or less according as they have been handled by men. Though I cannot prove it, I am confident that the Stock Pigeon and the Common Pigeon would breed together if they were paired : for the difference is not so. great between our little Domestic Pigeon and the Stock Pigeon, as between it and the large rough-footed or Roman Pigeon, with which however it breeds. Besides, in this species we can trace all the gradations between the wild and the domestic state, as they occur in succession; in the order of genealogy, or rather of degeneracy. The Stock Pigeon is imitated, in a way that cannot be misunderstood, by those deserters which leave our pigeon-houses; they perch on trees, which is the first and strongest shade in their return to the state of nature : these Pigeons, though bred domestics, and apparently reconciled, like the rest, to a fixed abode and to common habits, abandon their dwelling, renounce society, and seek a settlement in the woods; and thus, impelled by in-

stinct

ftinct alone, they refume their native manners.
Others, feemingly lefs courageous and lefs intrepid, but equally fond of liberty, fly from our
pigeon-houfes, and feek a folitary lodgment in the
holes of old walls, or, forming a fmall body, they
haunt fome unfrequented towers; and in fpite
of the hardfhips to which their fituation expofes
them, and the multiplied dangers that affail them
from all fides, they ftill prefer thefe uncomfort-
able dwellings to the convenience and plenty of
their former manfion: this is the fecond gradation to the ftate of nature. The Wall Pigeons
do not completely adopt their native habits, and
do not perch like the former, yet they enjoy a
much larger fhare of freedom than thofe which
remain in the domeftic condition. The third
gradation is the inhabitants of our pigeon-houfes,
which never leave their dwelling but to fettle in
one more comfortable, and which roam abroad
only to feek amufement, or to procure fubfiftence. And as even among thefe there are fome
deferters, it would feem that the traces of their
primæval inftincts are not entirely effaced. The
fourth and fifth gradations have totally changed
their nature. Their tribes, varieties, and inter-
mixtures are innumerable, being completely domeftic from the earlieft ages; and man, while he
has improved their external forms, has changed
their internal qualities, and extinguifhed in them
every fentiment of freedom. Thefe birds are
for the moft part larger and more beautiful than
the

the Common Pigeons; are more prolific, fatter and finer flavoured, and on all these accounts more pains have been bestowed upon them. They are inactive helpless creatures, that require the constant attention of man; and the most cruel hunger cannot in them call forth those little arts in which animals are usually so prompt. They are therefore completely domesticated, and entirely dependent on man, who has degraded them from their original condition.

If we suppose, that after our dove-cots were stocked, we selected those of the young which were most remarkable for their beauty, and raised them apart with greater care and attention, and still continued to choose the most gaudy of their descendants; we should at last obtain those painted varieties which at present exist. To give a complete history of these would therefore be to detail the effects of art, rather than to describe the productions of nature. For this reason, we shall content ourselves with the bare enumeration of them.

The BISET or WILD PIGEON *, is the primitive stock whence all the others are descended. It is commonly of the same size and shape with the

---

* *Columba-Livia,* Gesner, Gmel. and Briss.
Le Biset, Buff.
Columba Saxatilis, Aldrov.
Columba Fera Saxatilis, Schwenckfeld.
The Biset Pigeon, Lath.

Domesti

Domeſtic Pigeon, but of a browner colour. It varies however both in its bulk and plumage; for the one which is figured by Friſch under the name of *Columba Agreſtis*, is the ſame bird with a white ſhade, and its head and tail reddiſh; and what the ſame author has termed *Vinago*, *five Columba Montana*, is ſtill the Wild Pigeon, only its plumage borders on a dark blue. What Albin deſcribes by the term *ring-dove*, which is not applicable to it, muſt be conſidered as ſtill the ſame bird; and likewiſe what Belon calls the *Deſerter Pigeon*, which is more proper. We may ſuppoſe that this variety has ariſen from thoſe individuals which deſert our pigeon-houſes, and relapſe into the ſtate of nature; for the dark blue Wild Pigeons neſtle not only in the clefts of trees, but in the holes of ruins and precipices which they find in the foreſts. Hence ſome naturaliſts have called them *Rock Pigeons*, and others, becauſe they are fond of elevated tracts, have named them *Mountain Pigeons*. We may alſo obſerve, that this is the only ſpecies of the Wild Pigeon with which the ancients ſeem to have been acquainted, and which they called Οινας, or *Vinago*, and that they never mention our brown ſort, which is however the only Pigeon really wild, and never reduced to that ſtate of domeſtication. My opinion on this ſubject derives additional force from this fact, that in all countries where there are Domeſtic

Pigeons,

Pigeons, the *Oenas* is found, from Sweden * to
the torrid zone †; but the Brown Pigeon never
occurs except in cold regions, and continues
only during the fummer in our temperate climates. They arrive in flocks in Burgundy,
Champagne, and other northern provinces in
France, about the end of February and the beginning of March; they fettle in the woods and
neftle in the hollow trees, laying two or three
eggs in the fpring, and probably making a fecond
hatch in fummer; they raife only two young at
a time, and leave the country in November, and
direct their courfe towards the fouth, traverfing
Spain, to pafs the winter probably in Africa.

The Bifet or Wild Pigeon and the Oenas or
Deferter Pigeon, which returns into the wild
ftate, perch, and by this circumftance, they are

* " Cærulean Dove with a fhining neck, and a double blackifh
" fpot on the wings." LINN. *Fauna Suecica*, No. 174.

† " Wild and tame Pigeons are found every where in Perfia,
" but the wild ones are much the moft numerous; and as
" Pigeon's dung is the beft for melons, a great many Pigeons are
" carefully bred throughout the kingdom, and no country in the
" world, has, I fuppofe, more beautiful pigeon-houfes. . . . .
" Above 3000 pigeon-houfes are computed in the neighbourhood
" of Ifpahan; it is a pleafure to fee people take Pigeons in the
" field, by means of Pigeons tamed and trained for the pur-
" pofe, which they make to fly in flocks the whole day befide the
" Wild Pigeons; thefe are thus mingled in the flock, and led to
" the pigeon-houfe." *Voyage de* CHARDIN, tom ii. p. 29, and 30.
TAVERNIER, tom. ii. p. 22, and 23. " The Pigeons of the ifland
" Rodrigue are rather fmaller than ours, all of them flate coloured,
" and conftantly very fat and excellent: they perch and neftle on
" the trees, and are very eafily caught." *Voyage de* LEGUAT,
tom. i. p. 106.

distin-

diftinguifhed from the Wall Pigeons, which alfo
forfake their houfes, but feem afraid to penetrate
into the forefts.   After thefe three Pigeons, the
two laft of which approach more or lefs to the
ftate of nature, we fhall range the Common
Pigeon *, which, as we have obferved, is only
half domeftic, and ftill retains the original inftinct of flying in flocks.   If it has loft that native courage which is founded on the feeling of
independence, it has acquired more of the agreeable and ufeful qualities.   It often hatches thrice
a-year, and, if ftill more domefticated, even ten
or twelve times; whereas the Brown Pigeons
breeds only once, or at moft twice, annually.
They lay, at intervals of two days, almoft always
two eggs and feldom three, and never raife
more than two young, which are commonly a
male and female.   Many, and thefe are of the
younger fort, lay only once a-year, and the
fpring hatch is always the moft numerous.   The
beft pigeon-houfes are thofe built facing the
eaft, on fome rifing ground feveral hundred
paces diftant from the farm-yard ; where the inhabitants can enjoy quiet, have the advantage of
an extenfive profpect, and receive the cheering
influence of the morning fun.   I have frequently
feen Pigeons, flying from the vallies before fun-

---

* In Greek, περιστερα; in Latin, *Columba*; in Italian, *Colombo*,
or *Colomba* ; in Spanifh, *Colont*, or *Paloma* ; in German, *Taube*, or
Tauben; in Saxony, *Duve*; in Swedifh, *Duwa* ; and in Polifh,
Golab.

rife,

rife, alight to bafk on a pigeon-houfe that was feated on a hill, and drive away or even difpoffefs the lodgers; and this happens oftenefl in fpring and autumn. I fhall add another remark, that lofty and folitary pigeon-houfes are the moft productive. From one of mine, I had ufually 400 pairs of young Pigeons; while I got only 100 or 130 from others that were fituated 200 feet lower. The only danger is, left the rapacious birds that hover about the elevated tracts difturb the Pigeons and check their breeding, for they cannot much diminifh their numbers, as they prey on thofe only which ftray from the flock.

After the Common Pigeon, which is half do-meftic, we fhall place thofe varieties to the pro-duction of which man has fo much contributed; but the number is fo immenfe, that it would ex-ceed the limits of our work to defcribe each particularly, and we fhall therefore be contented with a general furvey.

The curious in this line apply the name of Bifet to all Pigeons that live in the fields, or are bred in large pigeon-houfes, and call thofe *do-meftic* which are lodged in fmall pigeon-houfes, or voleries, and do not venture to roam abroad. They are of different fizes: for inftance, the tumbler and wheeler Pigeons, which are the leaft of all the volery Pigeons, and fmaller than the Common Pigeon. They are more agile and nimble, and when they breed with the common fort, they lofe their diftinctive qualities. It

would

would feem that their peculiar conftrained motions are owing to the flavery to which they are reduced.

The pure breeds; that is, the principal varieties of the Domeftic Pigeons, from which all the fecondary ones can be derived; are: 1. The Pouter Pigeons *, which are fo called on account of their power of inflating their craw in refpiration; 2. The Proud Pigeons †, which are noted for their prolific quality, fuch as the Roman Pigeons, the rough-footed Pigeons, and the Jacobine Pigeons; 3. The Shaker ‡ Pigeons, which difplay their broad tail, like the turkey and peacock; 4. The Turbet Pigeon §, 5. The Shell Pigeon of Holland ‖; 6. The Swallow Pigeon ¶; 7. The Carmelite Pigeon **; 8. The Dafhed Pigeon ††; 9. The Swifs Pigeons ‡‡; 10. The Tumbler Pigeon §§; 11. The Wheeler Pigeon ‖‖.

The breed of the Pouter Pigeon confifts of the following varieties:

1. The Wine-fop Pouter Pigeon, in which the males are extremely beautiful, being decora-

---

* *Les groffes gorges,* i. e. the thick throats.

† *Les Pigeons mondains.*

‡ *Les Pigeons paons,* i. e. the Peacock Pigeons.

§ *Le Pigeon cravate,* ou *à gorge frifée*; i. e. the cravated or frizled-necked.

‖ *Le Pigeon coquille Hollandois.*　　　　¶ *Le Pigeon-hirondelle.*
** *Le Pigeon carme.*　　　　　　　　　†† *Le Pigeon heurté.*
‡‡ *Les Pigeons Suiffes.*　　　　　　　§§ *Le Pigeon culbutant.*
‖‖ *Le Pigeon tournant.*

ted

THE POUTER PIGEON .

THE POUTER PIGEON.

ted with plumage of a varied intermixture of hues; but the females are deftitute of fuch ornament.

2. The Painted Chamois Pouter Pigeon; the female has not that rich affemblage of colours. To this variety we ought to refer the Pigeon pl. cxlvi. of Frifch, and which the Germans call *Kropftaube* *, or *Kroüper*, and to which that author has applied the epithet *Strumous Pigeon*, or *Pigeon with the inflated œfophagus*.

3. The Pouter Pigeon, white as a Swan.

4. The White Pouter Pigeon, rough-footed, with long wings which crofs over the tail, and of which the ball of the neck appears very loofe.

5. The variegated gray, and foft gray Pouter Pigeon, whofe colour is delieate, and fpread uniformly over the whole body.

6. The Pouter Pigeon of iron gray, and barred, and ftriped gray.

7. The Gray Pouter Pigeon fpangled with filver.

8. The Hyacinth Pouter Pigeon, of a blue colour interwoven with white.

9. The Fire-coloured Pouter Pigeon; each of its feathers is marked with a blue and red bar, and terminates in a black bar.

10. The Hazel-coloured Pouter Pigeon.

11. The Chefnut-coloured Pouter Pigeon, whofe tail-quills are all white.

---

* i. e. *The crop or craw Pigeon.*

12. The

12. The Dark Pouter Pigeon of a fine velvet black, with ten wing-quills white, as in the Chefnut Pouter Pigeon. Both have the bib or kerchief under the neck white; and the females are like the males. Of all the Pouter Pigeons of a pure breed, that is, which have an uniform plumage, the ten quills are all white as far as the middle of the wing, and this may be regarded as a general character.

13. The Slaty Pouter Pigeon, which has the under-furface of the wings white, and a white cravat; the female is like the male.—Thefe are the principal breeds of the Pouter Pigeons, but there are others of inferior beauty, fuch as the red, the olive, the fable, &c

All Pigeons have more or lefs the power of inflating their craw by infpiring air; and the fame effect may be produced by blowing into the gullet. But this breed of Pouter Pigeons poffeffes the property in fo fuperior a degree as can refult only from fome peculiar conformation of its organs. The craw, almoft as large as the reft of the body, and kept conftantly inflated, obliges them to draw back their head, and prevents them from looking forward: and thus while they fwell with conceit, the falcon feizes them unawares. Hence they are raifed more for curiofity than utility.

Another breed is the Proud Pigeons; they are the moft common, and at the fame time the moft efteemed, on account of their prolific quality.

The

The Proud Pigeon is nearly one half larger than the Bifet, and the female pretty much resembles the male. They breed almost every month in the year, provided that only a small number are put into the same volery, and to each there be allowed three or four baskets or rather holes, formed into pretty deep casements with shelves, to prevent them from seeing one another while sitting; for each Pigeon not only defends its own hole, and fights the others that come near it, but contends for the possession of the next row. For example, eight pairs are sufficient to stock a space eight feet square, and people who have bred them affirm that six pair would be equally productive. The more their number be increased in a given space, the more there will be of brawling and fighting and of broken eggs. In this breed there are often impotent males, and barren females, which never lay.

They are fit to breed in the eighth or ninth month, but do not attain maturity till three years old. Their prolific powers are vigorous for six or seven years; after which the number of eggs they lay diminishes gradually; though there are instances of their breeding at the age of twelve. They lay their two eggs sometimes in the space of twenty-four hours, and during winter in that of two days; so that the interval varies according to the season. The female keeps her first egg warm, without covering it assiduously, nor does she begin to sit closely till after

the

the fecond is laid. The period of incubation is commonly eighteen days; fometimes only feventeen, efpecially in fummer, and nineteen or twenty in winter. The attachment of the female to her eggs is fo ardent and fteady, that fhe will forego every comfort, and fubmit to the moft cruel hardfhips, rather than forfake them. A Hen Pigeon, whofe toes froze and dropt off, perfifted to fit till her young were hatched: her toes were froft-bitten, becaufe her hole chanced to be clofe to the window of the dove-cot.

While the female is employed in hatching, the male places himfelf in the next hole; and the moment fhe is compelled by hunger to leave her eggs and go to the trough, he obferves her feeble murmur of intimation, takes her place, covers the eggs and fits two or three hours. This incubation of the male is commonly repeated twice in the courfe of the twenty-four hours.

The varieties of the Proud Pigeon may be reduced to three with refpect to fize, which have all the common character of a red filet round the eyes.

1. Thofe heavy birds that are nearly as large as fmall pullets; their bulk alone recommends them, for they are not good breeders.

2. The *Bagadais* are large Proud Pigeons, with a tubercle over the bill in the form of a fmall morel *, and a broad red ribbon round the

---

* *Morel* is a little red mufhroom.

eyes,

eyes, that is, a fecond eyelid, flefhy and reddifh, which even falls upon the eyes when they are old, and prevents them from feeing.—Thefe Pigeons are not productive.

The *Bagadais* have a curved and hooked bill, and exhibit many varieties; white, black, red, tawny, &c.

3. The Spanifh Pigeon; which is as large as a hen, and exceedingly beautiful. It differs from the *Bagadais* in not having the *morel* above its bill, and its fecond flefhy eyelid being lefs protuberant, and its bill ftraight inftead of curved. It croffes with the *Bagadais*, and produces a very thick and large breed.

4. The Turkifh Pigeon; which, like the *Bagadais*, has a thick excrefcence above the bill, with a red bar extending from the bill round the eyes. This bird is very thick, crefted, low legged, with a broad body and wings: fome are of a tawny colour, or a brown bordering on black, fuch as reprefented in Pl. CXLIX. of Frifch; others are of an iron-gray, lint-gray, chamois, and wine-fop. Thefe Pigeons are very inactive, and never roam from their volery.

5. The Roman Pigeons; which are not quite fo large as the Turkifh, but have the fame extent of wings, but no creft; they are black, tawny, or fpotted.

Thefe are the largeft of the Domeftic Pigeons; there are fome of a middle fize, and others fmaller. Among the rough-legged Pigeons, which are feathered as low as the nails, we may diftinguifh the one without a creft, figured by Frifch, Pl. CXLV. under the name *trummel taube* *, in the German ; *Columba typanifans*, in the Latin; and *Pigeon-tambour*, in the French : alfo the crefted rough-legged Pigeon, which the fame Author has defigned in Pl. CXLIV. by the name of *Montaube* † in German, and in Latin by the epithets *Columba menftrua, feu criftata pedibus plumofis* ‡. The rough-legged Drum-Pigeon is alfo termed the *Glou-glou Pigeon*, becaufe it continually repeats that found, and its voice at a diftance refembles the beat of a drum. The crefted rough-legged Pigeon is alfo called the *Month Pigeon*, becaufe it hatches every month, and does not wait till its young are able to provide for themfelves. Its breed is very profitable, though we muft not reckon upon twelve hatches annually ; the ufual number is eight or nine, which is ftill very great.

In the intermediate and fmall breed of Domeftic Pigeons, we may diftinguifh the Jacobine Pigeon §, of which there are many varieties; viz. the Wine fop, the Painted Red, and the Painted Chamois ; but in none of the three is the female thus decorated. In the Jacobine

---

* i. e. *The Drum Pigeon.* The Latin and French fignify the fame.

† i. e. *The Month Pigeon.*

‡ i. e. *The Month Pigeon, or the crefted with feathery feet.*

§ *Pigeon Nonain,* i. e. The Nun Pigeon.

breed,

THE JACOBINE PIGEON.

breed, there is alfo the *Moorish Pigeon* *; which is entirely black, except the head and the tips of its wings, which are white : to this we may refer the Pigeon of Pl. CL. of Frifch, which he names in German *Schleyer*, or *Parruquen-taube* †, and in Latin *Columba-galerita*, that is, Hooded Pigeon. But in general all of the Jacobines are hooded, or rather have a half-cowl on the head, which defcends along the neck, and extends along the breaft like a cravat of ruffled feathers : this variety is nearly allied to the Pouter Pigeon, for its fize is the fame, it alfo fomewhat inflates its craw, nor is it fo prolific as the other Jacobines, of which the moft perfect are entirely white. In all of them, the bill is very fhort ; the latter breed often, but their young is very fmall.

The Shaker or Peacock Pigeon is fomewhat larger than the Jacobine. The fineft of this breed have thirty-two feathers in the tail, while the common fort have only twelve. After they have raifed their tail, they bend it forwards, and at the fame time draw back the head fo as to make it meet the tail. They fhake alfo during the whole of this movement ; either from the violent contraction of the mufcles, or from fome other caufe, for there is more than one breed of Shaker Pigeons ‡. They make this difplay of the tail commonly

---

* *Pigeon Maurin.* † i. e. *The veiled* or *perruked Pigeon.*
‡ There is a Shaker Pigeon different from the Peacock Pigeon,

its

commonly in the love feafon; though fometimes alfo upon other occafions.   The female
raifes and difplays her tail alfo like the male, and
is quite as beautiful; fome kinds are entirely
white, others white with the head and tail black.
To this fecond variety we muft refer the Pigeon
figured in Pl. CLI. of Frifch, which he calls in
German, *Pfau-taube*, or *Hunerfchwantz* *, and
in Latin, *Columba caudata*.   That Author remarks at the fame time, that the Shaker Pigeon
difplays its tail, and works eagerly and conftantly with its head and tail, nearly in the fame way
as the *wryneck*.   Thefe Pigeons do not fly fo
well as the others; their broad tail catches the
wind, and they often fall to the ground; for
this reafon they are bred chiefly from curiofity.
However, thefe Pigeons, though by themfelves
they could perform no diftant journies, have
been carried into remote countries: in the
Philippine iflands, fays Gemelli Carreri, are
Pigeons that elevate and fpread their tail like
the peacock.

The Polifh Pigeons are larger than the Shaker
Pigeons.   Their diftinguifhing character is a
very thick and fhort bill, their eyes bordered

its tail not being near fo broad.   The Peacock Pigeon has been
denominated by Willughby and Ray, *Columba tremula laticauda*
(broad-tailed Shaker Pigeon); and the Shaker Pigeon *Columba
tremula angufticauda feu acuticauda* (narrow-tailed or fharp-tailed
Shaker Pigeon): the latter, though it does not raife or difplay its
tail trembles, they fay, almoft continually.

* i. e. *Peacock* or *Hen-tailed Pigeon.*

THE SHAKER PIGEON.

THE POLISH PIGEON..

THE TURBET PIGEON.

with a red circle, their legs very low.   They are
of different colours, many black, others rufous,
chamois, dotted with gray, or entirely white.

The Turbet Pigeon is one of the ſmalleſt,
being ſcarcely larger than a turtle, with which it
breeds.   The Turbet Pigeon is diſtinguiſhed
from the Jacobine, the former not having the
half cowl on the head and neck, but only a tuft
of feathers that appear to ruffle on the breaſt and
under the throat.   Theſe Pigeons are very hand-
ſome, well-made, and have a neat air ; ſome are
of the colour of wine-ſop, others chamois,
painted, rufous, gray, entirely white or black,
and others white with black mantles.   The laſt
variety is what Friſch repreſents in his CXLVII.
Plate, under the German name *Mowchen*, and
the Latin deſignation *Columba collo birſuto* *.   This
Pigeon has an averſion to pairing with other
Pigeons, and is not very prolific : it is beſides
very ſmall, and eaſily falls a prey to the rapacious tribes.   Upon all theſe accounts it is ſcarcely
ever raiſed.

The Pigeons called Dutch-ſhell Pigeons, becauſe on the back of the head are reverſed
feathers forming a ſort of ſhell, are alſo ſmall.
Their head is black, the end of the tail and the
tip of the wings are alſo black ; but all the reſt
of the body is white.   Some are red-headed,
blue-headed,   or the head and tail yellow ;

* Pigeon with ſhaggy neck.

the

the tail is ufually of the fame colour with the
head, but the wings are always white.　The firft
variety which has a black head, refembles fo much
the Sea-fwallow, that fome perfons have applied
to it that name; and with the more appearance of
analogy, as this Pigeon has not its body round
like moft of the reft, but long and very flender.

Befides the Shell Pigeon which we have juft
mentioned, there are other Pigeons which have
the head and tail blue; others where thefe parts
are black; others where they are red; others
where they are yellow: but in all the four the
extremity of the tail is of the fame colour with
thehead. They are nearly as large as the Peacock
Pigeons, and their plumage is very neat and
fingular.

There are fome named Swallow Pigeons, that
are not larger than turtles, and like them are
flender fhaped and of very nimble flight: the
whole of the under-fide of their body is white,
and the upper-fide, as well as the neck, the head
and the tail, black, or red, or blue, or yellow,
with a fmall cafque of the fame colours on the
head, but the under-fide of the head is always
white and fo is that of the neck.　To this variety we muft refer the *Galeated Pigeon* of Johnfton and Willughby, of which the principal
character is, that the feathers of the head and
thofe of the tail and the quills of the wings are
always of the fame colour, and the body of a
different colour; for example, the body white,

and

and the head, the tail, and the wings black, or
of some other colour, whatever it be.

The Carmelite Pigeon, which forms a different breed, is perhaps the lowest and the smallest
of all our Pigeons; it appears squatted like the
goat-sucker; it is also very rough-legged, the
feathers on its thighs being exceeding long, and
its legs remarkably short. The males and females
resemble each other, as in most of the other
breeds. It includes four varieties, which like
those of the preceding sorts, are also of an iron-
gray, chamois, wine-sop, and soft gray: but in
them all, the under-side of the body and of the
wings is white, all the upper-side of the body
being of the colours we have mentioned. Their
bill is smaller than that of a turtle, and they have
a little tuft behind the head, which draws to a
point as in the crested lark.

The Drum Pigeon or *glou-glou*, of which we
have spoken, is also very low and rough-legged,
but larger than the Carmelite Pigeon, and nearly
of the size of the Polish Pigeon.

The Dashed Pigeon, which is marked by a
daub, as it were, of a black, a yellow, or a
red pencil, above the bill only, and as far as the
middle of the head, with the tail of the same
colour, and all the rest of the body white, is
highly valued by the curious. It is not rough-
legged; it is of the size of the ordinary Proud
Pigeons.

The

The Swifs Pigeons are fmaller than the Common Pigeons, and not larger than the Bifets; they even fly as nimbly. There are feveral kinds of them; viz. thofe garnifhed with red, with blue, and with yellow, on a filky white ground with a collar, which forms a horfe-fhoe on the breaft, and is of an embroidered red. They have often two bars on the wings, of the fame colour with that of the horfe-fhoe.

There are other Swifs Pigeons not garnifhed with intermingled tints, fhaded over the whole body with an uniform flate colour, and without any collar or horfe-fhoe. Others are called *jafpered yellow collars*, *mailed yellow collars*, and others *very mailed yellow collars*, &c. becaufe they have collars of that colour.

There is ftill another variety of the Swifs Pigeons, called the *Azure Pigeon*, becaufe its plumage inclines more to blue than the preceding.

The Tumbler Pigeon is one of the fmalleft kind; that which Frifch has figured, Pl. CXLVIII. under the names *Tummel-taube*, *tumler*, *Columba geftuofa feu gefticularia*, is of a rufous brown; but fome are gray, and variegated with rufous and gray. It whirls round in its flight, like a body thrown in the air; for this reafon it has received its appellation. All thefe motions feem to imply vertigoes, which, as I have obferved, may be afcribed to the effect of domeftication. It flies very

fwiftly,

fwiftly, and foars higher than any; but its movements are precipitate and very irregular. Frifoh fayᶜ, that its fluttering refembles in fome meafure the capers of a rope-dancer; it has been called the Harlequin Pigeon (*Columba geftuofa*). Its fhape is pretty much like that of the Bifet; it is commonly employed to attract Pigeons from other dove-cots, becaufe it flies higher and farther, and continues longer on the wing than the reft, and more eafily efcapes the hawk.

The fame may be faid of the Wheeler Pigeon, which Briffon has called after Willughby, the *Smiter Pigeon* *; it turns round in its flight, and flaps fo vigoroufly with its wings, as to make as much noife as a mill-clapper; and often in the violence of its exertions, which feem to be almoft convulfive, it breaks fome of its wing-quills. Thefe Wheeler or Smiter Pigeons are commonly gray, with black fpots on the wings.

I fhall barely mention fome other varieties that are uncertain or fecondary, noticed by the nomenclators, and which belong undoubtedly to the breeds that we have defcribed, but to which, from the imperfect accounts given, we cannot refer them with accuracy or certainty.

1. The Norway Pigeon mentioned by Schwenckfeld, which is white as fnow, and which is probably a crefted rough-legged Pigeon, bigger than the reft.

---

* Columba Percuffor, *Will*. and *Briff*.

2. The

2. The Pigeon of Crete, according to Aldrovandus, or of Barbary, according to Willughby *; which has a very fhort bill, its eyes encircled with a broad ring of naked fkin, and its plumage blueifh, and marked with two blackifh fpots on each wing.

3. The frizzled Pigeon † of Schwenckfeld ‡ and Aldrovandus §, which is entirely white, and frizzled all over its body.

4. The Carrier Pigeon of Willughby ‖, which is much like the Turkifh Pigeon both by its plumage which is brown, and by its eyes which are encircled with a naked fkin, and its noftrils covered with a thick membrane. Thefe Pigeons, it is faid, were ufually employed to carry letters fpeedily to a diftance, when difpatch was needed, which gave occafion to the name.

5. The Horfeman Pigeon of Willughby ¶ and Albin, produced, they fay, by croffing the Pouter Pigeon and the Carrier Pigeon, and partaking of the qualities of both ; for it has the power of inflating its craw, like the Pouter Pigeon, and, like the Carrier Pigeon, its noftrils are covered with thick membranes. But it is probable that any other Pigeon might be trained to carry light matters, or rather to fetch them from a diftance : We need only feparate them from

---

* Columba Barbarica, feu Numidica, *Will.*
† The *Laced Pigeon*, Lath.   ‡ *Columba Crispa*, Schwenck.
§ *Columba crispis pennis*, Aldrov.
‖ *Columba Tabellaria*, Will.   ¶ *Columba Equis*, Will.

their

their female, and carry them to the place from whence the news is to be brought, and they will certainly return to their mate as foon as they are fet at liberty *.

Thefe five families of Pigeons are only, we fee, fecondary varieties of the firft, which we have defcribed from the obfervations of fome curious people, who have paffed their lives in breeding Pigeons, and particularly the Sienr Fournier, who has for feveral years had the charge of the voleries and poultry-yards of his Highnefs the Count of Clermont. That prince, who difcovered an early tafte for the arts, directed all forts of domeftic fowls to be collected from every quarter, and continually intermixed. In this way, from the Hen Pigeon alone, an amazing variety was produced entirely new, and yet bearing the impreffions of their original fpecies, though all furpaffing it in beauty.

---

* " In the pigeon-houfes of Cairo, fome males are feparated " from their females, and fent into the cities from which they wifh " to receive news: The meffage is written on a fmall bit of pa- " per, which is folded and then covered with wax; this is ftuck " under the wing of the male Pigeon, and in the morning after a " hearty meal, he is difmiffed, and proceeds ftraight to the dove- " cot where his female refides. He travels farther in one day, " than a man on foot could in fix." Pietro della Valle, tom. i. p. 416, & 417.

At Aleppo, Pigeons are employed to carry letters from Alexandretta to Aleppo, which they perform in lefs than fix hours, though the diftance is at leaft twenty-two leagues.

Voyage de THEVENOT, tom. ii. p. 73.

Tame

Tame Pigeons were known in ancient
Greece; for Ariſtotle ſays, that they hatch ten
or eleven times a year, and thoſe of Egypt
twelve times *. However, we may ſuppoſe that
large dove-cots where Pigeons breed only twice
or thrice annually, were not very common in
the time of that philoſopher. He divides the
genus into four ſpecies †; to wit, the Ring-
Pigeon, the Turtle, the Biſet, and the Common
Pigeon ‡; and it is the laſt which he mentions
as breeding ten times a-year. But this rapid
multiplication is found only in ſome of thoſe
that are highly domeſticated. Ariſtotle takes no
notice of the varieties of the tame Pigeons. Perhaps they were then few in number; but in
the time of Pliny they ſeem to have been greatly multiplied; for that naturaliſt mentions a large
breed of Pigeons that exiſted in Campania, and
tells us, that there were ſome curious perſons who
gave an extravagant price for a pair, whoſe pe-
digree could be traced, and that theſe were kept
in little turrets erected on the houſe-tops §. All
that the ancients have ſaid with reſpect to the inſtincts and habits of Pigeons, muſt be applied to
the domeſtic ſort, rather than to the inhabitants

* *Hiſtoria Animalium*, lib. vi. 4.   † *Hiſt. Anim.* lib. viii. 3.

‡ In the original, φαῆα or φασσα, πελεια or πελειας, τρυγων,
οκας or φαψ.

§ Lib. x. 37.—The purchaſe was made by Lucius Axius, before
Pompey's civil war, for the ſum of four hundred *denarii*, about
fifteen pounds ſterling; a price much higher than is given by bird-
fanciers at preſent.

of

of pigeon-houfes, which ought to be confidered
as an intermediaterace between the tame Pigeons
and the wild, partaking of the qualities of both.
They are all fond of fociety, attached to their
companions, and faithful to their mates ; a neat-
nefs, and ftill more the art of acquiring the graces,
befpeak the defire to pleafe; thofe tender careffes,
thofe gentle movements, thofe timid kiffes which
grow clofe and rapturous in the moment of blifs;
that delicious moment foon renewed by the return of the fame appetites and by the gradual fwell
of the foothing melting paffion ; a flame always
conftant, and ardor continually durable ; an un-
diminifhed vigour for enjoyment; no caprice,
no difguft, no quarrel to difturb the domeftic
harmony, their whole time devoted to love and
progeny ; the laborious duties mutually fhared;
the male affifting his mate in hatching and guarding the young :—If man would copy, what mo-
dels for imitation! [A]

[A] Specific character of the Common Pigeon, *Columba Do-
meftica* :—" It is cinereous, its rump white, there is a ftripe on its
" wings, the tip of its tail is blackifh." Linnæus reckons up
twenty varieties. 1. The Bifet, *Columba Livia* : 2. The Rock-
Pigeon, *Columba Saxatilis* : 3. The Roman Pigeon, *Columba Hi-
fpanica* : 4. The rough-footed Pigeon, *Columba Dafypus* : 5. The
Crefted Pigeon, *Columba Criftata* : 6. The Norway Pigeon, *Columba Norwegica* : 7. The Barbary Pigeon, *Columba Barbarica* :
8. The Jacobine, *Columba Cucullata* : 9. The Frizzled Pigeon,
Columba Crifpa : 10. The Turbit Pigeon, *Columba Turbita* :
11. The Peacock Pigeon, *Columba Laticauda* : 12. The Tumbler
Pigeon, *Columba Gyratrix* : 13. The Helmet Pigeon, *Columba
Galeata* : 14. The Turkifh Pigeon, *Columba Turcica* : 15. The
Carrier Pigeon, *Columba Tabellaria* : 16. The Cropper Pigeon,
Columba

Columba Gutturosa. 17. The Horseman Pigeon, *Columba Eques:* 18. The Smiter Pigeon, *Columba Percussor:* 19. The Turner Pigeon, *Columba Jubata:* 20. The Spot Pigeon, *Columba Maculata.*

Though Linnæus reckons the Biset a variety of the *Columba Domestica,* it is evidently the same with our Wood Pigeon, which he denominates *Columba Oenas,* and thus characterizes : " Cine- " reous, neck glossy green; stripe on the wings, and the tip of the " tail, blackish." In English, it bears the name of Stock Dove, being supposed to be the only original of all the domestic kinds. Multitudes of Stock Doves breed in the rabbit burrows on the downs of Suffolk, and the young are every year sold by the shepherds. The Rock Pigeons, as our Author observes, are the same birds : they are frequent in the South of Russia, and breed in turrets, and on the steep banks and rivers : in winter, vast numbers resort to the cliffs of the Orkneys.

FOREIGN BIRDS,

WHICH ARE RELATED TO THE PIGEONS.

FEW fpecies are fo generally fpread as thofe of the Pigeons; for having a very powerful wing and a well-fupported flight, they can eafily perform very diftant journies. Accordingly, moft of our wild and tame forts occur in every climate ; houfe Pigeons are bred from Egypt to Norway, and though they thrive the beft in warm countries, they fucceed alfo in the cold when care is taken. What proves that in general they are little affected by heat or cold, is that the Wild Pigeon is almoft equally diffufed through the whole extent of both continents *.

* The birds which the inhabitants of our American iflands call *Wood Pigeons, (Ramiers,)* are the real European Bifets. They are migratory, and never halt long in one place. They follow the crops which ripen not at the fame time in all the different parts of the iflands. They perch on the talleft trees, in which they breed twice or thrice a-year. . . . . It is incredible what number the fportfmen kill. When they eat good grain, they are very fat and as well tafted as the Pigeons of Europe ; but thofe which feed on bitter feeds, fuch as thofe of the Acomas, are as bitter as foot. Du Tertre, Hift. Antilles, tom. ii. 256.—" There are Pigeons " on the coaft of Guinea, which are the moft common, fuch as our " Field Pigeons, and which are very good eating." Bosman's *Voyage to Guinea.*—There are many Pigeons in the Maldive iflands . . . . . At Calcutta are very large Pigeons and wild Peacocks. *Voyage de* Pyrard, p. 131 and 426.

The

The Brown Pigeon of New Spain, mentioned
by Fernandez under the Mexican name *Cehoilotl**,
which is entirely brown excepting the breaſt and
the tip of the wings which are white, appears to
be only a variety of the Biſet.   Its eyes are
encircled by a bright red ſkin; its iris black;
its legs red.   The one mentioned by the ſame
author under the name *Hoilotl* †, which is
brown marked with black ſpots, is probably but
a variety of the preceding, occaſioned by difference of age or ſex.   Another of the ſame country, termed *Kacahoilotl* ‡, which is blue in the
upper parts, and red on the breaſt and belly, is
perhaps only a variety of our Wild Pigeon.   All
theſe ſeem to belong to our European Pigeon.

The Pigeon deſcribed by Briſſon by the name
of *Violet Pigeon of Martinico* §, and which he
figures under this ſame name, appears to us only
a very ſlight variety of the Common Pigeon.
The one which that author calls ſimply the *Martinico Pigeon* ‖, and which is deſigned in our

* *Hiſt. Nov. Hiſp.* cap. cxxxii. It is the *Columba Mexicana* of
Briſſon and Gmelin ; and the *Mexican Pigeon* of Latham.

† *Ibidem*, cap. lvi. and lx. It is the *Columba Nœvia* of Gmelin,
the *Oenas Mexicana* of Briſſon, and the *Black-ſpotted Pigeon* of
Latham.

‡ *Ibidem*, cap. clix. The *Columba Cœrulea* of Gmelin, and the
Columba Cœrulea Mexicana of Briſſon, and the *Blue Pigeon* of
Latham.

§ " The Violet-cheſnut Pigeon; its belly tawny; its quill-fea-
" thers rufous within."   B R I S S.

‖ " Pigeon, duſky-rufous above, dilute fulvous wine-coloured
" below; the neck gold-violet; black ſpots on either wing; the
" lateral tail quills furniſhed with a blaek ſtripe, white at the
" tips."   B R I S S.

Planches

Planches Enluminées under this appellation of *Ru-fous Cayenne Pigeon*, are neither of them different from the Common Pigeon. It is probable even that the latter is the female of the former, and that both derive their defcent from the Deferter Pigeons. They are improperly called *Partridges* in Martinico, where no real partridges exift; but there are Pigeons that refemble partridges in colour only, and differ not confiderably from our European Pigeons.

As the one was brought from Cayenne and the other from Martinico, we may infer that they are fpread through all the warm countries in the New Continent.

The Pigeon defcribed and figured by Edwards (Pl. CLXXVI). under the name of the *Brown Indian Dove*, is of the fame fize with the Bifet, and as it differs only in colour, we may confider it as a variety produced by the influence of climate. Its eyes are encircled by a fine blue fkin devoid of feathers, and frequently it raifes its tail of a fudden, but does not however difplay it like the Peacock-pigeon.

In like manner, Catefby's *Paſſenger Pigeon* *, which Frifch terms the *American Pigeon*, differs
from

* This is the *Columba Migratoria* of Linnæus, and the Wild Pigeon of Lawfon. Its fpecific charaćter :—" Its orbits bare and " blood-coloured, its breaſt rufous." Weight nine ounces. The Paſſenger Pigeons breed in the northern parts of the American continent : they neſtle on trees, and lays two eggs. During incubation, they live on the feeds of the red-maple, and afterwards on thofe

from thofe which defert our pigeon-houfes, and
relapfe into the ftate of nature, only by the colours, and by the greater length of the tail-feathers, which feem to indicate an analogy with
the turtle: but thefe differences are too minute
to form a diftinct and feparate fpecies.

The fame may be faid of the Pigeon noticed
by Ray, called by the Englifh the *Parrot-
Pigeon*, afterwards defcribed by Briffon, and
which we have caufed to be delineated in the
Planches Enluminées, No. 138. by the name,
Green Pigeon of the Philippines : it differs from
our Wild Pigeon only by the intenfity of its
colours, which we may attribute to the effect of
a hot climate.—In the Royal Cabinet, there is a
bird termed the *Green Amboyna Pigeon*, which is
different from that on which Briffon beftows the
fame name. It is figured No. 163. *Planches
Enluminées*, and fo nearly refembles the preceding, that it may be confidered as a variety of

of the elm.   As foon as their provifions fail, they gather in vaft
bodies, and advance towards the fouthern provinces.   In hard
winters the air is darkened by their flight; one flock fucceeds to
another, and this paffage lafts feveral days.   When they rooft in
trees, the branches are fometimes broken down by their weight,
and the ground beneath is covered to a confiderable depth with
their dung.   The Indians ufed to kill vaft numbers, and collected
their fat to ferve as butter.   In the State of New York, the Paf-
fenger Pigeons are obferved in their progrefs to the fouthern or
weftern fettlements about the beginning of Auguft, and in their return about the beginning of March: they fly in mornings and
evenings.   Prodigious quantities of thefe birds are caught in clap-
nets, or decoyed and fhot.   T.

the

THE WHITE BELLIED PIGEON.

the fame, refulting from a difference of age or
of fex.

The Green Amboyna Pigeon defcribed by
Briffon * is of the fize of a turtle, and though different in the diftribution of its colours from that to
which we have appropriated that name, muft ftill
be confidered as but another variety of the European Pigeon. It is alfo extremely probable that
the Green Pigeon from the ifland of St. Thomas
mentioned by Marcgrave, which is of the fame
fize and fhape ,with the European fort, but differs from it, as from all others, by the faffron
colour of its legs, is alfo a variety only of the
Wild Pigeon. In general, Pigeons have all red
legs; the difference confifts entirely in the intenfity or the vivacity of this colour, and perhaps
the yellow caft obferved by Marcgrave was occafioned by fome diftemper or accident. It is
much like the Green Pigeons of Amboyna, and
of the Philippines, as delineated in the *Planches
Enluminées.* Thevenot fpeaks of thefe Green
Pigeons in the following terms: " In India, at
" Agra, there are found Green Pigeons, which
" differ from ours only in the colour. Fowlers
" catch them eafily with bird-lime."

The Jamaica Pigeon, mentioned by Sir Hans
Sloane †, which is of a purple brown on the

* " Olive-green Pigeon; the back chefnut; the wing-quills
" black above, cinereous below, their exterior margins yellow;
" the feet naked." BRISSON.

† " The leffer white-bellied Pigeon." SLOANE. —The middle
white-bellied Pigeon, BROWN.

body,

body, and white under the belly, and nearly of the fame fize with our Wild Pigeon, muft be regarded as merely a variety of that fpecies, efpecially as it is not a conftant inhabitant of Jamaica, but only vifits the ifland as a bird of paffage.

There is ftill another in Jamaica, which muft alfo be regarded as a variety of the Wild Pigeon; it is what Sloane, and after him Catefby, termed the White Crowned Pigeon : its fize is the fame ; it neftles and breeds alfo in the holes of rocks, fo that we can fcarce doubt of its being the fame fpecies.

From this enumeration it appears, that the Wild Pigeon of Europe is found in Mexico, Martinico, Cayenne, Carolina, and Jamaica ; that is, in all the warm and temperate climates of the Weft Indies; and that it alfo occurs in the Eaft, from Amboyna to the Philippines.

THE RINGDOVE.

The RING PIGEON*.

Le Ramier, Buff.
Columba Palumbus, Linn. and Gmel.
Palumbus, Gefner and Briffon.
Palumbus Torquatus, Aldrov. Ray and Will.
Columba Torquata, Frifch.
The Ring Dove, *Queeft*, or *Cufhat*, Will. and Penn.

As this bird is much larger than the Bifet, and as both are nearly related to the Domeftic Pigeon, we may fuppofe that the fmall breeds of our houfe-pigeons have proceeded from the Bifets, and the large breeds from the Ring Pigeons: and this conjecture is the more probable, as the ancients were in the practice of rearing and fattening the Ring Pigeons †. The only circumftance that feems to oppofe this idea is, that the fmall domeftic Pigeons crofs with the large forts, while the Ring Pigeon feems not to intermix with the Bifet, and, though they inhabit the fame tracts, do not mix together. The turtle, as it is ftill more eafily raifed and kept

---

* In Greek, Φασσα or φατ]α: in Latin *Palumbes* or *Palumbus*: in Italian, *Colombo Torquato*, *Colombaccio*: in Spanifh, *Paloma Torcatz*: in German, *Ringel Taube*: in Swifs, *Schlag-tub*: in Dutch, *Ring-duve*: in Flemifh, *Kriefs-duve*: and in the Brabant, *Manfeau*: in Swedifh, *Ring-dufwa*: and in the ifle of Oeland, *Siutut*: in Danifh, *Ringel-due*: and at Bornholm, *Skude*: in Polifh, *Grywatz*.

† Perrottus, *apud Gefnerum.*

in

in houfes, might equally be regarded as the
fource of fome of our domeftic breeds, were it
not, like the Ring Pigeon, of a peculiar fpecies,
that intermingles not with the Wild Pigeons.
But though in their native forefts, where each
can poffefs its proper female, thefe birds are never
obferved to affociate together, yet when they are
deprived of their liberty, and have no longer the
opportunity of felection, the force of paffion
may obliterate the principle of choice, and may
impel them to unite with the females of their
kindred fpecies, and give birth to a progeny of
hybrids. Nor will the offspring, like the males,
be blafted with fterility, but may, like the breed
between the he-goat and the fheep, be capable of
reproduction. To judge from analogy, the Pigeon
tribe confifts in the ftate of nature of three principal fpecies, as we have obferved, and of two
that may be regarded as intermediate. On thefe
the Greeks beftowed five different names : the
firft and largeft is the *Phaffa* or *Phatta*, which is
our Ring Pigeon : the fecond is *Peleias*, which
is our Wild Pigeon : the third is the *Trugon* or
the Turtle : the fourth, which is the firft of the
intermediate kinds, is the *Oinas*, which being
rather larger than the Wild Pigeon, muft be confidered as relapfed from the ftate of domeftication : the fifth is the *Phaps*, which is a Ring
Pigeon fmaller than[*] the *Phaffa*, and for that
reafon called the *Leffer Pigeon*, but which appears to us to be only a variety of the fpecies of
the

the Ring Pigeon; for it has been obferved that the Ring Pigeons are of different fizes, according to the nature of the climate.—Thus all the nominal fpecies, ancient or modern, may be reduced to three, viz. the Wild Pigeon, the Ring Pigeon, and the Turtle, which have all perhaps contributed to the endlefs varieties of our domeftic Pigeons.

The Ring Pigeons arrive in our provinces in the fpring, rather earlier than the Bifets, and retire in autumn fomewhat later. The month of Auguft is the time in France when the young are the moft numerous; and it appears that they iffue from the fecond hatch, which is made towards the end of the fummer; for the firft hatch being very early in the fpring, the neft is not covered with leaves, and is therefore too much expofed and often deftroyed. Some Ring Pigeons remain in moft of our provinces during winter; they perch like the Bifets, but do not like them conftruct their nefts in holes of trees; they place them on the tops, and build them neatly with fticks: the neft is flat and fo wide as to admit at once both the male and female. I have afcertained that very early in the fpring, they lay two and often three eggs; for feveral nefts have been brought to me containing two and fometimes three young ones already ftrong in the beginning of April *. Some perfons have

afferted

---

* Salerne fays, that the poulterers of Orleans buy, in the feafon of the nefts, a confiderable number of Turtles, which they blow

with

afferted that in our climate they breed only
once a-year, unlefs they are robbed of their eggs
or young, which, it is well known, obliges all
birds to a fecond hatch. But Frifch affirms that
they lay twice a-year, which feems to us very
certain; fince the union of the male and female
being conftant and faithful, would feem to imply that their love, and the attachment for
their young, continues the whole year. But the
female lays a fortnight after the embrace of the
male *, and fits only another fortnight ; and the
fame length of time would be fufficient for the
young gaining ftrength to enable them to fly,
and provide for themfelves :—thus it is probable, that fhe may breed twice in the courfe of
the year, firft in the beginning of fpring, and
again at the fummer folftice, as the ancients remarked. In warm and temperate climates this

with the mouth and fatten with millet, fo that in lefs than a fortnight they are fit to be carried to Paris. That in the fame
manner they fatten the Ring Pigeons, and alfo carry thither
Bifets and other Pigeons, which they call *Poftes,* and which
are, according to them, Pigeons that have forfaken dove-cots, and
roam at will, neftling fometimes in one place, and fometimes in
another, in churches, in towers, in the walls of old caftles, or in
rocks.—This fact proves that the Ring Pigeon, like all the Pigeons
and Turtles, can be reared like other domeftic birds, and confequently that they may have given birth to the moft beautiful and
the largeft dove-cot Pigeons. M. Le Roy, Lieutenant of the chaces,
and infpector of the park at Verfailles, affures me, that the young
Ring Pigeons taken from the neft, tamed and fattened very well,
and that even the old Ring Pigeons caught by a net, were eafily
reconciled to live in the voleries, where in a very fhort time by
blowing they grow fat.

* Ariftotle, *Hift. Anim.* lib. vi. 4.

undoubtedly

undoubtedly takes place, and very probably the cafe is nearly the fame in cold countries. The Ring Pigeon has a louder fort of cooing than the Common Pigeon, but is never heard except in the love feafon, and in fine weather; for when it rains, thefe birds are filent, and feldom does their fong cheer the gloom of winter. They live upon wild fruits, acorns, beech-maft, ftrawberries, of which they are very fond, and alfo beans and grain of all kinds. They make great havoc among the corn when it is fhed or lodged, and if thefe forts of food fail them, they have recourfe to herbage. They drink like other Pigeons, that is at one draught, without raifing their head, till they have fwallowed as much water as they have occafion for. As their flefh, efpecially when they are young, is excellent meat, their nefts are much fought for, and great numbers are robbed. This devaftation, joined to their flow multiplication, much reduces every where the fpecies. Many are caught indeed with nets in their route through the provinces bordering on the Pyrenees; but this lafts only a few days and at one feafon.

It appears that though the Ring Pigeons prefer the warm and temperate climates *, "they

* The rocks of the two iflands of Magdalena ferve as a retreat to an infinite number of Ring Pigeons, natives of the country, and differing not from thofe of Europe, except that they are of a more delicate and exquifite flavour. *Voyage au Senegal,* par M. ADANSON.

alfo

alfo inhabit fometimes the bleak regions of the
north; fince Linnæus inferts them among the
birds that are natives of Sweden *. They would
feem alfo to have migrated from the one continent to the other †; for we have received from
the fouthern parts of America, as well as from
the hot countries in the Old World, feveral birds,
which muft be confidered as varieties or fpecies
clofely allied to the Ring Pigeon, and which we
fhall notice in the following article. [A]

* *Fauna Suecica*, No. 175.

† At Guadaloupe the feeds of the logwood-tree, which were ripe,
had attracted a prodigious number of Ring Pigeons; for thefe birds
are paffionately fond of fuch feeds. They fatten furprifingly, and
their flefh acquires a very agreeable odour of cloves and nutmegs.
When thefe birds are fat they are exceffively lazy. . . . Several
difcharges of a mufket will not force them to rife; they only hop
from branch to branch, while they behold their companions drop
around them. *Nouveau Voyage aux îles de l'Amerique*, tom. v. p. 486.
In the Bay of All Saints, there are two forts of Ring Pigeons,
fome of the bulk of our Ring Pigeons, others fmaller, and of a
light gray : both are very good to eat, and the flocks of them are
fo large, that from the month of May to September, one man may
kill nine or ten dozen in a morning, when the fky is cloudy and
they refort to feed on the berries which grow in the forefts.
DAMPIER'S *Voyage*.

[A] Specific character of the Ring Pigeon, *Columba Palumbus* :—
" Its tail-quills are black behind, its primary wing-quills whitifh
" on their margin, the neck white on both fides."

FOREIGN BIRDS,
WHICH ARE RELATED TO THE RING PIGEON.

––––––––––

I.

THE Ring Pigeon of the Moluccas, mentioned under this name by Briſſon*, and which we have cauſed to be deſigned (*Pl. Enl.* No. 164.) with a nutmeg in its bill, becauſe it feeds on that fruit. How different foever the climate of thoſe iſlands be from that of Europe, the bird is ſo like our Ring Pigeon in ſize and figure, that we cannot but conſider it as a variety occaſioned by the influence of climate.

The ſame may be ſaid of the bird deſcribed by Edwards under the name of the *Triangular Spotted Pigeon* †, and which he tells us is found in the ſouthern parts of Guinea. As it is half rough-legged, and nearly of the ſize of the Eu-

––––––––––

* Columba Ænea, *Linn.* and *Gmel.* Palumbus Moluccenſis, *Briſſ.* The Nutmeg Pigeon, *Lath.*

Specific character:—" its legs feathery; its bill and legs " greeniſh ; its body copper-coloured."

† Columba Guinea, *Linn. Gmel.* and *Klein.* The Turtle of the Cape of Good Hope, *Sonnerat.*

Specific character : —" Its orbits naked and red; its wings marked " with triangular white ſpots; its tail-quills black at the tip."

ropean

ropean Ring Pigeon, we fhall refer it to that fpecies as a fimple variety. It differs indeed in its colours, being marked with triangular fpots on the wings, having the whole of the under-fide of the body gray, the eyes encircled with a red naked fkin, the iris of a fine yellow, the bill blackifh : but all thefe differences of the colour of the plumage, bill, and eyes, may be confider-ed as variations introduced by the climate.

A third variety of the Ring Pigeon, which occurs in the other continent, is the *Ring-tailed Pigeon* mentioned by Sir Hans Sloane and Brown *, which being nearly of the fame fize with the European fort, may be referred to it better than to any other fpecies. It is remark-able for the black bar which croffes its blue tail, for the iris, which is of a more lively red than in the Ring Pigeon, and for two tubercles near the bafe of its bill.

---

* " Pigeon with a ring-tail, or marked with a dufky belt." *Sloane.* Greater Pigeon, of a fky-black, the tail ftriped. *Brown.* Columba Caribæa. *Gmel.* The Ring tailed Pigeon. *Lath.*

II. The

II.

The FOUNINGO.

Columba Madagafcarienfis, Linn. and Gmel.
Palumbus Madagafcarienfis, Briff.
The Madagafcar Pigeon, Lath.

The bird called at Madagafcar *Founingo-mena-rabou*, and of which we retain part of the name, becaufe it appears to be a peculiar fpecies, and which, though related to the Ring Pigeon, differs too much from it in fize to be regarded as a fimple variety *. Briffon firft noticed this bird, and we have caufed it to be figured (*Pl. Enl.* No. 11.) under the appellation of the *Blue Ring Pigeon of Madagafcar*. It is much fmaller than the European Ring Pigeon, and nearly of the fame fize with another Pigeon of the fame climate, which appears to have been firft mentioned by Bontius †, and afterwards by Briffon ‡,

---

* What induces us to confider the Founingo of a different fpecies from our Ring Pigeon, is that the latter occurs in the fame climate. "We faw (fays Bontekoe), in the ifland of Mafcare-" nas, a number of Blue Ring Pigeons, which allowed themfelves " to be caught in the hand. We killed this day near two hun-" dred. . . . We alfo found there a number of Ring Pigeons." *Voyage aux Indes Orientales.*

† " Pigeon of a very green colour."

‡ Columba Madagafcarienfis, *Linn.* and *Gmel.* Palumbus Viridis Madagafcarienfis, *Briff.*

Specific character: " Its legs feathery; its tail violet; its body " bluifh-black."

from

from an individual brought from Madagafcar, where it was called *Founingo Maitfou;* which feems to prove that, notwithftanding the difference of colour, its being green inftead of blue, thefe two birds are of the fame fpecies, and the only diftinction fubfifting between them arifes from the age or fex. This bird is reprefented *Pl. Enl.* No. 111. by the name of *Green Ring Pigeon of Madagafcar.*

III.

The SCALLOP NECKED PIGEON.

Le Ramiret, Buff.
Columba Speciofa, Gmel.

We have reprefented this bird *Pl. Enl.* No. 213. by the name of the *Cayenne Ring Pigeon.* The fpecies is new, and has been defcribed by no preceding naturalift. It is fmaller than our Ring Pigeon, and different from the African *Founingo.* It is one of the handfomeft birds of this kind; it refembles fomewhat the turtle in the fhape of its neck, and the difpofition of its colours, but differs in point of fize, and in many other characters which denote a greater affinity to the Ring Pigeon, than to any other fpecies.

IV. The

IV.

The Pigeon of the Nincombar, or rather the Nicobar, iflands, defcribed and defigned by Albin *, which, according to him, is of the fize of the European Ring Pigeon: its head and throat are of a blueifh-black, the belly of a blackifh-brown, and the upper parts of the body and of the wings are variegated with blue, with red, with purple, with yellow, and with green. According to Edwards, who has, fince Albin, given an excellent defcription and an accurate figure of it, the fize does not exceed that of an ordinary Pigeon . . . . The feathers covering the tail are long and pointed like thofe of a dung-hill cock; they have very beautiful reflections of colour variegated with blue, with red, with gold, and with copper; the back and the upper-fide of the wing are green, with reflections of gold and copper . . . . I have, fubjoins Edwards, found in Albin, figures which he calls the *Cock* and the *Hen* of this fpecies; but I have examined the fpecimens in Sir Hans Sloane's collection, and can difcover no difference from which we might infer that thefe birds were male

---

* Columba Nicobarica, *Linn. Gmel.* and *Klein*. Columba Nicombarienfis, *Briff*. The Nicobar Pigeon, *Alb. Edw.* and *Lath*.

Specific character :—" Its tail is white, its body black, its " wing-quills blue, its back gloffy green, with an elongated feather " round its neck."

and

and female.   Albin calls it the *Ninckombar Pigeon;* the true name of the ifland whence this bird was brought is *Nicobar* . . . there are feveral fmall iflands which bear that name, and lie on the north of Sumatra.

V.

The bird called by the Dutch *Kron-vogel,* figured by Edwards Pl. CCCXXXVIII. under the name of the *Great Crowned Pigeon* *, and alfo by Briffon, by the term *Crowned Pheafant of India.*

Though this bird is as large as a turkey, it belongs undoubtedly to the genus of the Pigeon: its bill, its head, its neck, the general fhape of its body, its legs, its feet, its nails, its cooing, its inftincts, its habits, &c. all are analogous. From being deceived by its fize, and never thinking of comparing it with a Pigeon, Briffon, and afterwards our defigner, termed it a *Pheafant.* The laft work of Edwards was not then pub-lifhed ; that excellent ornitholgift has fince given

---

* Columba Coronata, *Linn.* and *Gmel.* Columba Mugiens, *Scop.*

Specific character :—" It is bluifh ; above cinereous ; its orbits
" black, its fhoulders ferruginous."

his

his opinion on the fubject. " It is of the family
" of the Pigeons, though it is as large as a
" middle fized turkey . . . Mr. Loten brought
" feveral of thefe birds alive from India . . . It
" is a native of the ifland of Banda . . . . Mr.
" Loten affured me that it was really a Pigeon,
" and has all the geftures and cooing of that bird
" in careffing its female : I confefs that without
" this information, I fhould never have imagined
" that a bird of fuch magnitude was related
" to the Pigeons *."

The Prince of Soubife has very lately received
at Paris, five of thefe birds alive. They are all
fo much like each other in fize and colour, that
it is impoffible to diftinguifh their fex. Befides,
they do not lay, and Mauduit, an intelligent naturalift, informs me, that he faw feveral in Holland, which alfo did not lay. I remember to
have read in fome voyages, that it is ufual in
India to raife thefe birds as we do our poultry.

* Edwards, Gleanings.

The COMMON TURTLE *.

La Tourterelle, Buff.
Columba Turtur, Linn. and Gmel.
Turtur, Gefner, Aldrov. Briff. Frifch, &c.
Palumbus-Turtur, Klein.
The Turtle-dove, Willughby.

THE Turtle, more perhaps than any other
bird, loves coolnefs in fummer, and
gentle warmth in winter. It arrives in our
climates very late in the fpring, and departs in
the end of Auguft; whereas the Bifets and the
Ring Pigeons appear a month earlier and remain
a month later, and fome even the whole winter.
All the Turtles, without a fingle exception, affemble in flocks, and perform their journeys in a
body; they never refide with us more than four
or five months, and, during that fhort fpace,
they pair, build their neft, and lay and rear their
young, which are able to join them in their
retreat. They choofe the darkeft and cooleft
woods to form their fettlement, and they conftruct their neft, which is almoft quite flat, on

* In Greek, Τρυγων, from τριζω or τρυζω, *to murmur :* the Latin
name *Turtur*, is evidently formed in imitation of the Turtle's notes
tur, tur; in Italian, *Tortora, Tortorella* ; in Spanifh, *Tortota* or
Tortora; in German, *Turtel, Turtel Taube*; in Swedifh, *Turtur
Dufwa*; in Polifh, *Trakawke*.

the

THE TURTLE.

the talleft trees at a diftance from our habitations. In Sweden *, in Germany, in France, in
Italy, in Greece †, and perhaps in countries
ftill cooler or hotter than thefe, they remain only during fummer, and depart before
autumn: only Ariftotle informs us, that in
Greece a few ftay behind in the moft fheltered fituations: this feems to prove that they
feek very hot climates where to pafs the winter. They are found in every part almoft of
the Ancient Continent ‡; they occur alfo in the
New,

* " The Turtles do not winter with us .... the Turtles keep
" in flocks when they arrive and depart .... the Quails alfo re-
" tire, except a few that fettle in fheltered fpots, which is likewife
" the cafe with the Turtles." Arist. *Hift. Anim.* lib. viii.

† " We faw in the kingdom of Siam, two forts of Turtles: the
" firft is like ours, and the flefh excellent; the fecond has a
" finer plumage, but its flefh is yellowifh and ill-tafted. The
" fields are full of thefe Turtles." *Second Voyage de Siam,* p. 248.
and Geronier, *Hift. Nat. and Polit. de Siam,* p. 35.—" Ring
" Pigeons and Turtles come to the Canary iflands from the coafts
" of Barbary." *Hift. Gen. des Voy.* tom. ii. 241.—At Fida in A-
frica, there is fuch a multitude of Turtles, that a man who fhot
pretty well, undertook to kill a hundred in fix hours time. Bosman's
Voyage to Guinea.—There are Turtles in the Philippines, in the
ifles of Pulo Condor, and in Sumatra. Dampier's *Voyage.*—Here
(at New Holland) is a number of plump fat Turtles, which are
very good eating. *Idem.*

" ‡ The plains of Chili are ftocked with an infinite number of
" birds, particularly Ring Pigeons, and Turtles." Frezier's
Voyage ... " The Ring Pigeons there are bitter, and the Turtles
" not excellent." *Idem.*—" In New Spain are many European birds,
" as Pigeons, large Turtles like thofe of Europe, and others as little
" as Thrufhes." Gemelli Carreri, tom. v.—" In no part of
" the world have I feen fuch numbers of Turtles and Ring Pigeons,

New *, as far as the South Sea iſlands †. They are, like the Pigeons, ſubject to varieties, and though naturally more ſavage, they can be raiſed in the ſame manner, and multiplied in the do-meſtic ſtate. It is eaſy to intermingle their different varieties, and they can even be made to breed with the Pigeon, and thus produce new

" as at Areca in Peru." LE GENTIL, tom. i. "In the country
" about the Bay of Campeachy, there are different ſorts of Turtles;
" ſome have a white craw, the reſt of the plumage gray verging
" on blue; theſe are the largeſt and are good eating; others are
" of a brown colour over the whole body, not ſo fat as the firſt,
" and ſmaller. Theſe two ſpecies fly in pairs, and live upon the
" berries which they gather from the trees. The third ſort are of a
" very dull gray, and called *Land Turtles*; they are much larger
" than a Lark, round and plump; they go in pairs." DAMPIER's
Voyage.—" It is commonly believed that there are Red Par-
" tridges and Ortolans at St. Domingo; but this is a miſtake,
" for theſe are different ſpecies of Turtles: ours are very common
" there." CHARLEVOIX, *Hiſt. des St. Dominque*, tom. i. pp. 28,
& 29.—" At Martinico and the Antilles, Turtles are ſeldom found
" but in ſequeſtered ſpots whither they are driven. Thoſe of Ame-
" rica have appeared to me to be much larger than thoſe of France·
" At the time they breed, many of the young are caught in nets;
" they are fed in voleries, and fatten perfectly well, but are not ſo
" fine taſted as the wild ones: it is impoſſible to tame them. Thoſe
" which live at liberty feed on *monbin* plums and wild olives, of
" which the nuts remain pretty long in the craw, which has led ſome
" perſons to believe they eat ſmall ſtones. They are commonly
" very fat and well taſted." *Nouv, Voy. aux îles de l'Amerique*,
tom. ii. p. 237.

* In the enchanting iſlands of the South Sea, we ſaw Turtles
that were ſo familiar as to perch upon us. *Hiſt. des Navig. aux
Terres Auſtrales*, tom. ii. p. 52. . . There are plenty of Turtles at
the Gallapago iſlands in the South Sea: they are ſo tame, that one
may kill five or ſix dozen in an afternoon merely with a ſtick.
Nouv. Voy. aux îles de l'Amerique, tom. ii. p. 67.

† Linnæus, *Fauna Suecica*, No. 175.

15   tribes

tribes, or new individual varieties. " I have
" feen, a perfon writes me of the moft un-
" doubted credit *, I have feen in Bugey, at a
" houfe of Chartreux, a bird got by croff-
" ing a Pigeon with a Turtle; it was of the
" colour of a French Turtle, and refembled the
" Turtle more than the Pigeon; it was reftlefs,
" and molefted the other birds of the volery.
" The father Pigeon was of a very fmall kind,
" perfectly white, with black wings." It has
not been obferved whether thefe hybrids are
prolific; but the general fact proves at leaft the
great analogy that fubfifts between thefe two
birds. It is therefore not unlikely, as we have
before remarked, that all the varieties of the domeftic Pigeon may refult from the gradations of
intercourfe, and the multiplied combinations of
the Bifet, the Ring Pigeon, and the Turtle.

What feems to confirm our opinion with regard to thefe unions, which may be conceived to
be illegitimate, as being out of the ufual courfe
of nature, is, that exceffive ardor which thefe
birds feel in the feafon of love. The Turtle melts
with a ftill more tender paffion than the Pigeon,
and more fingular preludes announce the fwell
of pleafure. The male Pigeon only ftruts round
his mate, puffing and difplaying his figure. The
Turtle, whether kept in confinement or fluttering
at will in the grove, begins his addreffes by

* M. Hebert, whom I have already cited more than once.

faluting

faluting his female eighteen or twenty times in fucceffion in the moft humble pofture, bending fo low each time as to touch the ground, or the branch, with his bill, and he fighs the tendereft murmurs. The female appears at firft infenfible to his paffion, but the fecret flame foon kindles, and at laft yielding to the foft defires, fhe gives vent to fome plaintive accents. And when once fhe has diffolved in his embrace, fhe burns with a conftant fire; fhe never leaves the male, fhe returns his kiffes and his careffes, and ftimulates him to renew the rapturous joys, till the bufinefs of hatching divides her attention, and invites to more ferious occupations.

I fhall cite only one fact which manifefts the ardour of thefe birds *: if the males be put in one cage and the females in another, they will copulate together as if they were of different fexes: the males indeed burn fooner and with more intenfity than the females. Confinement therefore only deranges nature, but cannot extinguifh it!

In the fpecies of the Turtle we are acquainted with two conftant varieties. The firft is, the

* The Turtle, M. Roy writes me, differs from the Ring Pigeon and the Common Pigeon, by its diffolutenefs and inconftancy, notwithftanding its reputation for the contrary qualities. Not only females that are fhut up in voleries receive promifcuoufly all the males; but I have feen wild ones, which were neither conftrained nor corrupted by domeftication, give favours to two fucceffively on the fame branch.

Common

THE COLLAR'D TURTLE.

THE WHITE TURTLE.

Common Turtle: the fecond the *Collared Tur-tle* *, fo called, becaufe it bears on its neck a fort of black collar.   Both of them are found in our climate, and when they mix together, they produce a hybrid. The one which Schwenckfeld defcribes, and which he calls *Turtur mixtus*, was the offspring of a common male Turtle, and of the female Collared Turtle, and refembled more the father than the mother. I have no doubt but thefe are prolific. The Collared Tur-tle is only fomewhat larger than the common kind; its inftincts and habits are the fame.   In general we may fay, that all the three tribes of the Pigeon are more analogous in their difpofi-tions than in their figure, They eat and drink in the fame manner, without lifting their head till they have fwallowed as much as they want; they fly in flocks; their voice is a loud mur-mur, or a plaintive moan, rather than an ar-ticulated fong; they lay only two eggs, fome-times three; all of them hatch feveral times in the year in warm countries, or when kept in voleries.   [A]

---

* Columba Riforia, *Linn.* and *Gmel.*   La Tourterelle à Collier, *Buff.*   Turtur Torquatus, *Briff.*   The Indian Turtle, *Albin,* & *Will.*   The Collared Turtle, *Lath.*

Specific character:—" Above yellowifh, with a black crefcent on " the neck."

[A] Specific character of the Turtle, *Columba-Turtur :*—" Its " tail-quills are tipped with white, its back gray, its breaft carna-" tion ; a black lateral fpot on its neck, with white ftrokes." The Turtle is found in the weft of England, where it breeds retired in the oak-woods.

FOREIGN BIRDS,

WHICH ARE RELATED TO THE TURTLE.

I.

Columba Marginata, Linn. and Gmel.
Turtur Americanus, Briſſ.
The Marginated Pigeon, Lath.

THE Turtle, as well as the Common Pigeon
and the Ring Pigeon, has ſuffered varieties
in different climates, and occurs likewiſe in both
continents. That which Briſſon has mentioned
by the name of the *Canada Turtle*, and which
is figured No. 176. *Pl. Enl.* is rather larger, and
its tail longer, than that of the European Turtle;
but the differences are not ſo great as to conſtitute a diſtinct ſpecies. I think that we might refer to it the bird which Edwards calls (Pl. XV.)
the *Long-tailed Dove*, and which Briſſon names
the *American Turtle*. Theſe birds much reſemble each other, and as they are diſtinguiſhed
from our Turtle only by the length of their
tail, we regard them as varieties produced by
influence of climate.

II. The

II.

The SENEGAL TURTLE and the COLLARED TURTLE OF SENEGAL, both mentioned by Briſſon, the ſecond being only a variety of the firſt, as the Collared Turtle of Europe is only a variety of the common ſort ; they appear not diſtinct ſpecies from our Turtles, for they are of the ſame ſize, and ſcarce differ but in the colours, which muſt be aſcribed to the influence of climate.

We preſume that the Spotted throated Turtle of Senegal, being of the ſame ſize and climate with the preceding, is alſo but a variety.

III.

The TOUROCCO.

Columba Macroura, Gmel.
The Great-tailed Pigeon, Lath.

But there is another bird of Senegal, which has hitherto been noticed by no naturaliſt, and which we have cauſed to be engraved *Pl. Enl.* No. 329. under the name of the *Broad-tail Turtle of Senegal*, this denomination being given

it

it by Adanſon when he preſented it.   However, as it ſeems to differ from the European Turtle, carrying its tail like the *Hocco*, and having the bill and other charaćters of the Turtle, the term *Tourocco* may denote its mixed qualities.   [A]

[A] Specific charaćter of the *Columba Macroura*:—" It is " cinnamon-coloured, below partly whitiſh, the tip of its tail " white."

IV.

The TURTLETTE.

Columba Capenſis, Gmel.
The Cape Pigeon, Lath.

Another bird a-kin to the Turtle; which is that deſcribed by Briſſon, and figured *Pl. Enl.* No. 140. by the appellation of *Black Cravated Turtle of the Cape of Good Hope*: but we have appropriated a name to it, becauſe it appears a peculiar ſpecies, different from that of the Turtle.   It is much ſmaller than our Turtle, and its tail much longer, though not ſo broad as that of the Tourocco : the two feathers in the middle of the tail only are very long.   The male alone is repreſented in the *Pl. Enl.*; it is diſtinguiſhed from the female by a kind of cravat of a ſhining black under the neck and on

the

the throat, while the correſponding part in the female, is gray mixed with brown.   This bird is found at Senegal, as well as at the Cape of Good Hope, and probably in all the ſouthern parts of Africa.   [A]

[A] Specific charaƈter of the *Columba Capenſis :—* " Its primary " wing-quills are rufous on the inſide."

V.

The   T U R V E R T.

We give this name to a green bird which bears ſome reſemblance to the Turtle, but appears to be a ſpecies entirely diſtinƈt from all the reſt. Under the Turvert we include three birds; No. 142, 214, and 117. of the *Pl. Enl.*   The firſt has been deſcribed by Briſſon, under the appellation of *Green Amboyna Turtle*, and in the *Pl. Enl.* by the *Purple-throated Turtle of Amboyna **, becauſe that colour of the throat is the moſt ſtriking charaƈter of the bird †.   The
                                    ſecond

* Columba Viridis, *Linn.* and *Gmel.* The Green Turtle, *Lath.* Specific charaƈter :—" It is copper-coloured, the under-ſide of its " body purple violet."

† To this ſpecies the following paſſages probably refer. " In " the iſland of Java, there is an infinite number of Turtles of dif- " ferent colours ; green with white and black ſpots ; yellow and " white, white and black, and a pecies of an aſh-colour. Their " bulk is as different as their colours are various : ſome are as large
                                    " as

fecond is the *Turtle of Batavia** , which has not been noticed by any naturalift. We may prefume that being a native of the fame climate with the Turvert, and differing little in fize, fhape, or colours, it is only a variety arifing from the age or fex. The third is termed the *Java Turtle* †, becaufe it is faid to inhabit that ifland; it feems alfo to be only a variety of the Turvert, but ftill more charaterifed than the former, by the difference of colour in the lower parts of the body.

VI.

Thefe are not the only fpecies or varieties of the Turtle tribe; for, in the Old Continent, we find the *Portugal Turtle* ‡, which is brown, with black and white fpots on each fide, and near the

" as a Pigeon, and others are fmaller than a Thrufh." Le GENTIL *Voyage au Tour du Monde.*

" In the Philippine iflands is a fort of Turtle which has the " feathers on the back gray, and thofe on the ftomach white; in " the middle of which we perceive a red fpot like a frefh wound " flowing with blood." GEMELLI CARRERI, tom. v. p. 266.

* Columba Melanocephala, *Gmel.* The Black-capped Pigeon, *Lath.*

† Columba Javanica, *Gmel.* The Javan Turtle, *Lath.*

‡ Columba Turtur, *Var.* 3. *Gmel.* Turtur Lufitanicus, *Briff.*

middle

middle of the tail: *The ſtriated Turtle of China* *, which is a beautiful bird, the head and neck being ſtreaked with yellow, red and white : *The ſtriated Turtle of India* †, which is not ſtriped longitudinally along the back as the preceding, but tranſverſely on the body and the wings: *The Amboyna Turtle* ‡, which is alſo ſtriped tranſ- verſely with black lines on the neck and breaſt, with a very long tail.   But as we have not ſeen theſe four birds, and as the authors who deſcribe them term them *Doves* or *Pigeons*, we cannot decide whether they belong to the Pigeons or to the Turtles.

---

* Columba Sinica, *Linn.* and *Gmel.* Turtur Sinenſis Striatus, *Briſſ.* Dove from China, *Alb.*
Specific charaɛter :—" It is duſky, ſtriped with black ; its belly " ſomewhat blood-coloured ; its wings yellow, the wing-quills " and the bill black.''

† Columba Striata, *Linn.* and *Gmel.* Turtur Indicus Striatus, *Briſſ.* The Barred Turtle, *Lath.*
Specific charaɛter :—" Its orbits and ſtraps bright white ; its body " cinereous, ſtriped with black, below rufous.''

‡ Columba Amboinenſis, *Linn.* and *Gmel.* Turtur Amboinenſis, *Briſſ.* Thus deſcribed by Briſſon, " Rufous; tail very long; neck " and breaſt covered with feathers ſtriated tranſverſely with black- " iſh ; wing-quills duſky ; tail-quills of a duſky-rufous.''

VII. The

VII.

The   T O U R T E.

Columba Carolinenſis, *Columba Canadenſis,*	Linn. and Gmel.
Turtur Carolinenſis, *Turtur Canadenſis,*	Briſſ.
The Carolina Pigeon, *The Canada Pigeon,*	Penn. and Lath.

In the New Continent we meet firſt with the Canada Turtle, which, as I have ſaid, is the ſame ſpecies with the European Turtle.

Another bird, which we have called after the travellers, *Tourte,* is what Cateſby has termed the *Carolina Turtle* *. It appears to be the ſame, the only difference being a gold-coloured ſpot, mixed with green and crimſon, which in Cateſby's bird is placed below the eyes and on the ſide of the neck, but which is not to be ſeen in ours. This would incline me to ſuppoſe that the firſt is the male, and the ſecond the female. It is likely that the *Picacuroba* of Brazil, mentioned by Marcgrave, belongs to this ſpecies.

I preſume alſo that the Jamaica Turtle †, noticed by Albin and afterwards by Briſſon, being

---

* This Pigeon reſides the whole year in Carolina, and feeds on the berries of poke (*Phytolacca Decandria,* Linn.) and the ſeeds of the mug-apple. *(Podophyllum Peltatum,* Linn.) Its fleſh is delicate.

† Columba Cyanocephala, *Linn.* and *Gmel.* Turtur Jamaicenſis, *Briſſ.* The Turtle Dove from Jamaica, *Alb.* The Blue-head Turtle, *Lath.*

Specific character :—" Its head is blue, with a white ſtripe under " its eyes."

a naa native of the fame climate with the preceding, and differing but little from it, muſt be regarded as a variety of it.

We ſhall alſo remark, that this bird bears a great reſemblance to the one given by Edwards, which is probably only the female of ours. What alone ſeems oppoſed to this opinion, is the difference between the climates.   Edwards was informed that his bird came from the Eaſt Indies, and ours was brought from America. Might not there be ſome miſtake with regard to the climate of Edwards's ? Theſe birds are ſo much like each other, and to the *Tourte*, that we cannot be perſuaded that they are the inhabitants of climates ſo widely different; and we are certain that ours was ſent from Jamaica to the Royal Cabinet.

VIII.

The C O C O T Z I N.

Columba Paſſerira, Linn. and Gmel.
Turtur Parvus Americanus, Briſſ.
Columbus Minutus, Klein.
The Ground Dove, Cateſby, Penn. and Lath.

We have retained this name given by Fernandez, becauſe the bird on which it was beſtowed ſeems to differ from all the others.   As it is ſmaller than the Ordinary Turtle, many naturaliſts have called it the *Little Turtle* *.   Others

---

* Ray, Sloane, Brown, &c.

have

have called it the *Ortolan* *, becaufe it is not much larger than that bird, and is excellent eating. It was reprefented *Pl. Enl.* No. 243; by the name of *Little Turtle of St. Domingo*, fig. 1. and *Little Turtle of Martinico*, fig. 2. But after a clofe examination and comparifon, we are convinced that they are the fame bird; fig. 2. being the male, and fig. 1. the female. It would alfo feem that the *Picuipinima* of Pifo and Marcgrave, and the Little Turtle of Acapulco, mentioned by Gemelli Carreri †, belong all to the fame kind. And thus this bird is fpread through all the fouthern parts of the New World. [A]

* Martinico Ortolan, *Dutertre.*—" The birds which our ifland-
" ers call *Ortolans*, are only Turtles much fmaller than thofe of
" Europe . . . Their plumage is of an afh-gray, the under-fide
" of the throat inclines fomewhat to rufous: they always go in
" pairs, and many of them are found in the woods. Thefe birds
" are fond of feeing people, and come into the roads without being
" fcared. When taken young, they grow very tame : they are
" lumps of fat of a lufcious tafte." *Nouv. Voy. aux îles de l'Ame-*
rique, tom. ii. p. 237.

† " In the neighbourhood of Acapulco, Turtles are feen fmaller
" than ours, with the tips of the wings coloured; they fly even
" into houfes." Gemelli Carreri, tom. vi. p. 9.

[A] Specific chara&er of the *Columba Pafferina:*—" The quills
" of its wings and tail are darkifh, its body is purplifh, its bill and
" legs are red." This Pigeon is not larger than a Lark. It
fometimes advances to the coaft of Carolina, where it feeds on the
berries of fhrubs, efpecially thofe of the pellitory.

END OF THE SECOND VOLUME.

Printed in the United States
By Bookmasters